Complete Course Notebook

Intermediate Algebra

FIFTH EDITION

Alan S. Tussy
Citrus College

R. David Gustafson
Rock Valley College

Prepared by

Ann K. Ostberg
Grace University

BROOKS/COLE
CENGAGE Learning

Australia • Brazil • Japan • Korea • Mexico • Singapore • Spain • United Kingdom • United States

For product information and technology assistance, contact us at
**Cengage Learning Customer & Sales Support,
1-800-354-9706**

For permission to use material from this text or product, submit all requests online at **www.cengage.com/permissions**
Further permissions questions can be emailed to
permissionrequest@cengage.com

ISBN-13: 978-1-133-36392-7
ISBN-10: 1-133-36392-X

Brooks/Cole
20 Davis Drive
Belmont, CA 94002-3098
USA

Cengage Learning is a leading provider of customized learning solutions with office locations around the globe, including Singapore, the United Kingdom, Australia, Mexico, Brazil, and Japan. Locate your local office at: **www.cengage.com/global**

Cengage Learning products are represented in Canada by Nelson Education, Ltd.

To learn more about Brooks/Cole, visit **www.cengage.com/brookscole**

Purchase any of our products at your local college store or at our preferred online store **www.cengagebrain.com**

Printed in the United States of America
1 2 3 4 5   16 15 14 13 12

FD088

Complete Course Notebook for Intermediate Algebra

Table of Contents

Complete Course Notebook Overview (for Instructors)

Dear Educators,

Cengage Learning is pleased that you have selected our Tussy series for your course. We are excited to present the **Complete Course Notebook** to be used in conjunction with your textbook.

Students entering our colleges today begin in developmental studies courses for a number of reasons. Research has shown that some of the reasons include inadequate knowledge base of subject area, weak study skills, test anxiety, lack of organization, poor note-taking skills, and /or low reading comprehension. By addressing these areas and improving study skills, the Complete Course Notebook will bolster student success in the journey through higher education.

Engaging students in an active learning process allows them to develop the skills necessary for success in the higher education classroom setting. Incorporating the Complete Course Notebook into your class addresses the aforementioned concerns in a way that will not be just "extra work" for students and extra grading for instructors. In order to achieve optimum results, the student should engage in the text's study skills workshops, which have been incorporated into this notebook, and in the notebook's section-specific tools, which will prepare them for class, help them organize what they learn in lectures, and allow them to practice the skills they learn after class. A detailed description of each section's tools is given in the **Complete Course Notebook Overview (Students).**

To fully benefit from the Complete Course Notebook, students should purchase the textbook, this notebook, a one- or two-inch three-ring binder with side pockets, and a set of dividers (one for introductory materials, each chapter in the textbook, and a few extra for tests and handouts). The Complete Course Notebook is perforated and three-hole punched so students can insert its pages into this three-ring binder and separate the sections with dividers as they go.

The Complete Course Notebook will help students develop organizational skills necessary for success in your classroom, future courses, the workforce, and life. We welcome your comments and suggestions about the new Complete Course Notebook, and we wish you the best of luck in the coming term.

Cengage Learning & the Tussy Algebra series

Complete Course Notebook Overview (for Students)

Dear Students,

Cengage Learning is pleased to present you with the **Complete Course Notebook** to be used in conjunction with your textbook. This notebook will guide you through essential study skills that will help you to be successful in this course and in your future studies. This notebook, most importantly, will not just be "extra work" for you, but will guide you through the essential work that needs to be done in order to ensure that you successfully complete your algebra course.

To benefit from the Complete Course Notebook, you will need the following:
- The textbook
- The Complete Course Notebook (which is perforated and three-hole punched)
- A one- or two-inch three-ring binder
- A set of dividers (you will need a divider for introductory materials and one for each chapter of the textbook, plus a few extra for tests and other papers from your instructor)

By tearing out the pages of this notebook and inserting them into a three-ring binder, you will be able to insert all the work you do here alongside tests, quizzes, and homework for that section.

How the Complete Course Notebook is Organized:
For each chapter and section of your textbook, this notebook contains a correlating chapter introduction and section-specific work. The chapters are organized as follows:

Study Skills Workshop: This adaptation of the book's Study Skills Workshop for that chapter will allow you to easily complete the activities associated with the workshops. The workshops focus on a variety of topics related to succeeding in your math course and beyond.

Pre-Class Prep: These prepare you to get the most out of your classroom experience. The three activities within the Pre-Class Prep should be completed prior to attending class.
  - ✓ *Are You Ready?* is the chapter readiness assessment located at the beginning of each section in your textbook. These quick problems review skills that will be needed in the current section.
  - ✓ *Reading Time!* is designed to increase your focus while reading the chapter and to provide exposure to the topics to be discussed in lecture. This easy-to-complete activity includes fill-in-the-blank, true/false, identification, and/or matching questions.
  - ✓ *Getting Ready for Class* allows you to identify concepts that look confusing/challenging and separate those from the concepts you are confident you understand.

In-Class Notes: As your instructor is presenting each section, use this two-column format to write examples and notes directly across from the terms, definition, and main ideas. Add any additional binder paper as needed.

After Class: These are filled with study skills that you will use after listening to a lecture, while completing an assignment, and after the completion of an assignment. These five activities will help to make your after-class experience a positive one.

- ✓ *Important to Know* focuses on having a specific place to write down your homework assignment and/or any other requirements from your instructor and due date(s).
- ✓ *Getting to Work!* focuses on recording when you have a question about a particular problem so that you know exactly what you would like to ask your instructor, tutor, or another student.
- ✓ *Do You Really Know It?* focuses on writing to explain the concepts being studied in such a way that you and others understand.
- ✓ *You Write the Test!* focuses on being able to critically evaluate a section's content as to what a student should be expected to know on a test. You will write one to two questions to include in *My Practice Chapter Test.*
- ✓ *Reflect on the Section* focuses on reflecting on the lecture to determine if you understand topics you identified prior to class as confusing or challenging.
- ✓ *Reflect on Your Math Attitude* focuses on your attitude in a variety of areas, how your attitude affects your performance, and suggestions to help you through the course. This activity will also incorporate the skills addressed in the Study Skills Workshops.

Chapter Activities: These one- to two-page activities may be assigned by your instructor to complete in class or on your own to solidify your understanding of chapter topics.

Chapter Test Skills Assessment: Suggestions are given for pre-test preparation and practice.

My Practice Chapter Test: This practice test is the compilation of the test questions and answers that you wrote for *You Write the Test!* in the *After Class* section. This practice test provides another way to prepare for your tests.

> **Note:** You may choose to take the pages out of the 3-ring binder during class while taking notes; however once the pages are filled in and re-inserted in the binder, you will be able to view each section's materials in a sequential format. After completing each homework assignment, you will be able to place it in your notebook right with the other pages that correspond to that particular section of the book.

In this first introduction section you will find:
- Peer Contact Information Guide
- Support System Worksheet
- Scheduling Your Work for This Course
- Note-Taking Suggestions
- Study Tips
- Test-Taking Strategies
- Helpful Websites and Resources

Best of luck for the upcoming term and in this course!

Cengage Learning & the Tussy Algebra series

Scheduling Your Work for This Course

Make copies of this blank calendar for each week of the course. Insert chronologically in your notebook.

This Week's Dates: _____ to _____

Top Priorities This Week:

- _____

- _____

Complete this calendar of what you need to accomplish in this class this week. Schedule these items on this calendar:

    ✓ *Read Section ____ and complete Pre-Class Prep*
    ✓ *Assignments to complete*
    ✓ *Complete Section ____ After Class*
    ✓ *Study for exams or quizzes*

Sunday __/__	Monday __/__	Tuesday __/__	Wednesday __/__	Thursday __/__	Friday __/__	Saturday __/__
	What is Due?	What is Due?	What is Due?	What is Due?	What is Due?	
Work to do:	Work to do:	Work to do:	Work to do:	Work to do:	Work to do:	Work to do:
Study Session Time: Where:	**Study Session Time:** Where:	**Study Session Time:** Where:	**Study Session Time:** Where:	**Study Session Time:** Where:	**Study Session Time:** Where:	**Study Session Time:** Where:
Time studied? __ hr ___ min	**Time studied?** __ hr ___ min	**Time studied?** __ hr ___ min	**Time studied?** __ hr ___ min	**Time studied?** __ hr ___ min	**Time studied?** __ hr ___ min	**Time studied?** __ hr ___ min

Total time studying this week: _____ hr _____ min   **Compare to *Target Goal*: above   below   just right**

If your hours are below your target goal, then you may have to reevaluate your schedule, especially if you are not performing well in the course; if your hours are above the target goal, visit with your instructor as you may be spending too much time "spinning your wheels." Let your instructor or tutor help you.

Support System Worksheet

You can gather the information you need to begin building your academic support system by answering the questions below.

1. What is your instructor's phone number and/or email address?

2. What are the days and times of your instructor's office hours?

3. Where does your instructor hold office hours?

4. Does your campus have a math tutoring center? If so, where is it located?

5. What is the phone number and email address for your math tutoring center?

6. What are the hours of operation for your math tutoring center?

7. Read the preface of your textbook. Does your textbook have any support systems such as videos, online tutoring, or textbook features that can help to learn on your own? List each resource below and where you can locate them.

Peer Contact Information Guide

It is important as you start your coursework each semester to share and gather contact information from peers in each class. By having this information, you will be able to contact others with questions about assignments or schedule a time to study together.

Name: _____

E-mail: _____

Social Network Account:
_____

Phone: _____

Preferred contact method:
_____

Name: _____

E-mail: _____

Social Network Account:
_____

Phone: _____

Preferred contact method:
_____

Name: _____

E-mail: _____

Social Network Account:
_____

Phone: _____

Preferred contact method:
_____

Name: _____

E-mail: _____

Social Network Account:
_____

Phone: _____

Preferred contact method:
_____

Name: _____

E-mail: _____

Social Network Account:
_____

Phone: _____

Preferred contact method:
_____

Name: _____

E-mail: _____

Social Network Account:
_____

Phone: _____

Preferred contact method:
_____

Note-Taking Suggestions

- Read the course material and make notes prior to coming to class. It is easier to engage in the learning process if you have done your *Pre-Class Prep* work.

- Date your notes and mark the section number of the book that correlates with the notes you are taking.

- Use different colored pencils/markers/highlighters. Use the margins to add arrows, stars, bullets, or specific feedback from your teacher.

- Listen for the main points and write down definitions and key vocabulary. It is better to write too much than not enough, but you don't want to write the lecture word-for-word.

- If your instructor indicates that something is important, make a note of it: star it, highlight it, or underline it.

- If your instructor is going too fast or is unclear, let him/her know. You will not be the only one feeling that way.

- You may choose to take a tape recorder to class if you cannot keep up with the pace of the lecture.

- Review your notes soon after you have taken them to fill in any holes you may have left.

- Place your notes in this three-ring binder with all your work in sequential order.

- Compare class notes with your peers, update yours, and allow them to do the same. Do this regularly; going over the material with another classmate helps you to process the information and to store it in your memory.

- Some colleges offer short courses in specific note-taking techniques. If you struggle to keep up and get the information recorded, look into one of these courses.

Study Tips

- You may wish to take a learning styles inventory quiz. The quiz will help you to understand how you learn and what methods are best for your note-taking and studying. Refer to Chapter 2 Study Skills Workshop *Preparing to Learn*.

- Study a little each day throughout the week. Try not to "cram" at the last minute.

- Take breaks as you study.

- Find a comfortable place with good lighting to study that is free of distractions.

- When you sit down to work, have all your materials together and available.

- You may want to study some alone and then study in a group setting to reinforce what you've learned. Verbalizing concepts helps you fully understand them. When choosing a study group, make sure you study with others who are serious about learning the material. Refer to Chapter 7 Study Skills Workshop *Study Groups*.

- The first information and the last information studied at each study session are usually retained best in your memory. Study the most important topics first and last.

- Work through the practice tests at the end of each chapter. Work the problems within a strict timeframe so you can learn to pace yourself. Do not forget the *My Practice Chapter Test* in this notebook at the end of each chapter.

- Study at the time of day when you learn best. Some people work best in the morning, others late at night.

- Attend class regularly. Teachers will often review prior to tests and some will hold study sessions to help prepare you.

- Make and use flashcards or a glossary often to review key ideas and definitions that may carry through the entire course.

Test-Taking Strategies

- Complete the Chapter 3 Study Skills Workshop *Successful Test Taking* in this notebook.
- Sit in the classroom away from distractions; front and center is often best.
- Arrive early with the necessary tools for the test (calculator, pencil, ruler...)
- Get plenty of sleep the night before.
- Eat something before testing, but not heavy foods.
- Use your time wisely.
- If you are unsure of how to solve a problem, skip it and come back to it later. If time permits, check over your work and answers.
- Read each question carefully to make sure you are giving the answer that is requested.
- Write legibly.
- Try to stay relaxed; if you are anxious, your brain is not working at its optimum level.
- Worrying about your performance will only hinder your work.
- Pace yourself so that you finish the exam.
- Underline important concepts as you are reading each question; small words can make a huge difference in the answer.
- Stay focused on the test.
- Dress appropriately so that you will not be too hot or cold.
- Come well-prepared.
- If you are unsure of the directions, ask the instructor to clarify.
- Write down formulas, definitions, or reminders to yourself somewhere on the test when you first receive it. Then you won't worry about getting confused or forgetting these specifics.
- A test is not a race. You have the entire period, so do not become frustrated when other students begin leaving the classroom.
- Have a good attitude and be confident that you've prepared yourself with your Complete Classroom Notebook!

Helpful Websites and Resources

Textbook Website:

www.cengage.com/math/tussy and www.cengagebrain.com

Learning Skills Quizzes:

A Learning Style Survey for College
      http://www.metamath.com/multiple/multiple_choice_questions.html

What's Your Learning Style? Quiz
      http://www.edutopia.org/mi-quiz

Math Videos:

If you use CourseMate or YouBook with this textbook, you will find free videos by Rena Petrello in each section to illustrate important section-specific topics.

Khan Academy
      http://www.khanacademy.org/
      Try the developmental math, algebra, and pre-algebra playlists. These videos are all free!

MathTV
      http://www.mathtv.com/videos_by_topic

Study Skills:
General Study Skills Resources
      http://www.howtostudy.org/

Study Environment Analysis
      http://faculty.bucks.edu/specpop/frame-ls-7.htm

Room for Instructors to Suggest Course-Specific and School-Specific Resources:

Chapter 1 A Review of Basic Algebra

Read the *Study Skills Workshop* found at the beginning of Chapter 1 in your textbook. **Complete** the activities below for this chapter's *Study Skills Workshop*.

Make Time for the Course

Each week, the *Scheduling Your Work for this Course* form (page 4 in this notebook) allows you to write down what you need to do that week. Managing your time well can make for a successful experience in this class.

✓ Complete a new *Scheduling Your Work for this Course* form every week.

A study time **Target Goal** is based on the guideline of 2 hours of independent study time for every hour in the classroom each week. If you are in class 3 hours per week, your **Target Goal** would be 6 hours.

✓ Determine your **Target Goal**: I should be spending approximately _____ hours in study time each week for this course.

On the *Scheduling Your Work for this Course* form, you will be able to evaluate the amount of time spent in study time. This information will allow you to make adjustments to your schedule.

Know What is Expected

From reading the syllabus, note important facts:

✓ Attendance policy: _____

✓ Homework: _____

✓ Exams/quizzes: _____

✓ Final: _____

✓ Late policy: _____

✓ Can/cannot use a calculator, if allowed, recommended calculator: _____

✓ My grade in this class will be based on _____

✓ Other class policies: _____

Build a Support System

✓ Complete the *Support System Worksheet* (page 5 in this notebook).
✓ Complete the *Peer Contact Information Guide* (page 6 in this notebook).

Are You Ready?

Complete the following problems. These review some basic arithmetic skills that are needed in this section. *Answers are found in the Pre-Class Prep Answer Section.*

1. What operation is indicated by each word?

   a. sum

   b. difference

   c. product

   d. quotient

2. If $m = 3d + 5$, find m if $d = 5$.

3. What is 22 increased by 10?

4. What is 6 less than 17?

 Reading Time!

While **reading Section 1.1**, fill in the blanks choosing from the following words (some may be used more than once or not at all). *Answers are found in the Pre-Class Prep Answer Section.*

sum	quotient	variables	words	two-column table
equation	mathematical	formula	difference	product
multiplying	expressions	dividing	subtracting	constant

1 Write Verbal and Mathematical Models

1. The result of multiplying is a _____. The result of adding is a _____.

2. The result of _____ is a quotient. The result of _____ is a difference.

3. A verbal model is written using _____.

4. Letters (or symbols) that stand for numbers are _____, while a number, such as 7, is an example of a _____.

5. Algebraic _____ are variables and/or numbers that are combined with the operations of addition, subtraction, multiplication, division, raising to a power, and finding a root.

6. Verbal models are translated into _____ models.

7. A mathematical sentence that contains an = symbol is an _____.

2 Use Equations to Construct Tables of Data

8. A _____ is an equation, which expresses a relationship between two or more variables.

9. When applications require the repeated use of a formula, the results can be shown in a_____.

✓ Getting Ready for Class

Briefly look through the section again. Answer the following by writing the concept or just the page number from the text.

Identify concepts/procedures that you feel confident about:

Identify concepts/procedures that look confusing or challenging:

Be sure to ask your instructor further questions if you are still having difficulty with a concept.

1.1 The Language of Algebra: In-Class Notes

Terms, Definitions, and Main Ideas	Examples and Notes

Use notebook paper for additional notes

1.1 The Language of Algebra: After Class

Important to Know

What is your homework assignment? Be sure to note it in your weekly schedule.

Section 1.1 Homework: _____ **Due**: _____

Getting to Work!

Complete your homework assignment. If you are unable to do a problem, write down the problem number and a question to help you remember what you would like to ask your instructor, your tutor, or another student.

Problem Number	Question? Where in the problem did you start to have difficulty or confusion?	Answered?

Often, you will have more questions than there is space provided here. If so, write them on notebook paper and be sure to talk to your instructor. You might ask in class or privately with the instructor.

Do You Really Know It?

Can you put into words the concepts that you learned in this section? Answer the below question from the *Writing* section in the *Study Set* in your text. Explain as if you were explaining to someone who has never taken this class before. Use notebook paper if you need more room.

Use each word below in a sentence that indicates a mathematical operation. If you are unsure of the meaning of a word, look it up in a dictionary.

quadrupled	deleted	bisected	garnished
confiscated	annexed	docked	quintupled

You Write the Test!

If you were writing the test for this section, what would you want a student to know? Write two test questions that you think might come from this material. Write questions of various difficulty; these questions can be original or chosen from the homework. Be sure to supply the answer also!

Write these questions at the end of this chapter under the section titled *My Practice Chapter 1 Test* and the answer to each question under the section titled *My Practice Chapter 1 Test Answers*.

Reflect on the Section

Look back at the *Pre-Class Prep* section. Did the lecture explain topics that you thought were going to be challenging or confusing? _____

- Are there topics that you still have questions on from the reading or the lecture? If so, complete the following:
    I don't understand…

- Speak to your instructor in class or during office hours about these concerns.

Reflect on Your Math Attitude

Write one word that describes how you feel about this class: _____

Was the word you wrote a positive word, a negative word, or a neutral word? Explain why.

To be successful, it is helpful to write "*I can do this*" on the top of your math papers. Read it and believe it. Keep a positive attitude. Believing that you have the ability to do the work will help you to be successful. Try writing "*I can do this*" on your next homework assignment!

1.2 The Real Numbers: Pre-Class Prep

 Are You Ready?

Complete the following problems, which review several types of numbers that we use in everyday life. *Answers are found in the Pre-Class Prep Answer Section.*

1. Count the number of letters in the word *supercalifragilisticexpialidocious.*

2. What number represents a temperature that is 27 degrees below zero?

3. What type of number is used to express a grade point average (GPA)?

4. Suppose a recipe calls for only part of a full cup of flour. What type of number is normally used to describe such an amount?

 Reading Time!

While **reading Section 1.2**, identify the statements as True or False. *Answers are found in the Pre-Class Prep Answer Section.*

1 Define the Set of Natural numbers, Whole Numbers, and Integers

_____ 1. Natural numbers ($\mathbb{N}$) are the set of numbers used for counting.

_____ 2. To write sets, list the elements of the set within braces { }.

_____ 3. The set of whole numbers (W) does not include 0.

_____ 4. $\mathbb{N} \subseteq W$ is read as "The set of whole numbers is a subset of the set of natural numbers."

_____ 5. A whole number greater than 1 that has only itself and 1 as factors is a composite number.

_____ 6. Two numbers that are the same distance from 0 on the number line, but on opposite sides of it, are called opposites.

_____ 7. Numbers to the left of 0 on a number line are called negative numbers.

_____ 8. The set of integers ($\mathbb{Z}$) is the set of whole numbers and their opposites combined.

_____ 9. Integers that are divisible by 2 are called even integers.

2 Define the Set of Rational Numbers

_____ 10. A rational number is any number that can be written in the form $\dfrac{a}{b}$, where a and b

represent integers.

_____ 11. If a and b represent numbers, where $b \neq 0$, then $\dfrac{a}{-b} = \dfrac{-a}{b} = -\dfrac{a}{b}$.

_____ 12. The set of rational numbers ($\mathbb{Q}$) can terminate or repeat in their decimal form.

_____ 13. Using set-builder notation, the set of rational numbers is

$$\left\{ \dfrac{a}{b} \middle| a \text{ and } b \text{ are integers, with } b \neq 0 \right\}.$$

3 Define the Set of Irrational Numbers

_____ 14. A irrational number cannot be expressed as a fraction.

_____ 15. The set of irrational numbers is denoted by the symbol $\mathbb{R}$.

_____ 16. An example of an irrational number is $\sqrt{400}$.

4 Classify Real Numbers

_____ 17. The set of real numbers ($\mathbb{R}$) is the combination of the set of rational numbers and the

set of irrational numbers.

_____ 18. A number can be a member of all the following sets: real numbers, integers, and

irrationals.

5 Graph Real Numbers

_____ 19. The coordinate, which is a real number, corresponds to a point on the number line.

_____ 20. To graph a number means to make a drawing that represents the number.

6 Order the Real Numbers

_____ 21. The inequality symbol < means "is less than."

_____ 22. The decimal equivalent of a fraction represents the same value.

7 Find the Opposite and the Absolute Value of a Real Number

_____ 23. The opposite of a number is the same as the additive identity.

_____ 24. The absolute value of a number is its distance from 0 on the number line.

_____ 25. For any real number a, if $a < 0$, then $|a| = a$.

Getting Ready for Class

Briefly look through the section again. Answer the following by writing the concept or just the page number from the text.

Identify concepts/procedures that you feel confident about:

Identify concepts/procedures that look confusing or challenging:

Be sure to ask your instructor further questions if you are still having difficulty with a concept.

Chapter 1 A Review of Basic Algebra

1.2 The Real Numbers: In-Class Notes

Terms, Definitions, and Main Ideas	Examples and Notes

Use notebook paper for additional notes

1.2 The Real Numbers: After Class

Important to Know

What is your homework assignment? Be sure to note it in your weekly schedule.

Section 1.2 Homework: _____ **Due**: _____

Getting to Work!

Complete your homework assignment. If you are unable to do a problem, write down the problem number and a question to help you remember what you would like to ask your instructor, your tutor, or another student.

Problem Number	Question? Where in the problem did you start to have difficulty or confusion?	Answered?

Often, you will have more questions than there is space provided here. If so, write them on notebook paper and be sure to talk to your instructor. You might ask in class or privately with the instructor.

Do You Really Know It?

Can you put into words the concepts that you learned in this section? Answer the below question from the *Writing* section in the *Study Set* in your text. Explain as if you were explaining to someone who has never taken this class before. Use notebook paper if you need more room.

Explain why every integer is a rational number, but not every rational number is an integer.

You Write the Test!

If you were writing the test for this section, what would you want a student to know? Write two test questions that you think might come from this material. Write questions of various difficulty; these questions can be original or chosen from the homework. Be sure to supply the answer also!

Write these questions at the end of this chapter under the section titled *My Practice Chapter 1 Test* and the answer to each question under the section titled *My Practice Chapter 1 Test Answers*.

Reflect on the Section

Look back at the *Pre-Class Prep* section. Did the lecture explain topics that you thought were going to be challenging or confusing? _____

- Are there topics that you still have questions on from the reading or the lecture? If so, complete the following:
    I don't understand…

- Speak to your instructor in class or during office hours about these concerns.

Reflect on Your Math Attitude

Think about what is expected from you in this course. You might want to review what you wrote in this chapter's *Study Skills Workshop* under *Know What is Expected*. Identify and elaborate on any concerns that you might have about these expectations.

Communicate these concerns with your instructor. Be sure to ask questions on anything that you do not understand or is not clear.

1.3 Operations with Real Numbers: Pre-Class Prep

 Are You Ready?

Complete the following problems. These problems review some basic concepts that are important when adding, subtracting, multiplying and dividing positive and negative real numbers. *Answers are found in the Pre-Class Prep Answer Section.*

1. Find $|4|$ and $|-6|$. Which number, 4 or -6, has the larger absolute value?

2. Add: $9.37 + 2.8$

3. What is the opposite of -11?

4. Subtract: $\dfrac{4}{9} - \dfrac{1}{6}$

5. Multiply: $\dfrac{14}{3} \cdot \dfrac{2}{21}$

6. Divide: $5.95 \div 0.7$

 Reading Time!
(on the next page)

 Getting Ready for Class

Briefly look through the section again. Answer the following by writing the concept or just the page number from the text.

Identify concepts/procedures that you feel confident about:

Identify concepts/procedures that look confusing or challenging:

Be sure to ask your instructor further questions if you are still having difficulty with a concept.

Reading Time!

While **reading Section 1.3**, match the word or concept to its definition or description. Not all choices are used. *Answers are found in the Pre-Class Prep Answer Section.*

1 Add and Subtract Real Numbers

_____ 1. Sum

_____ 2. Difference

_____ 3. Add two positive numbers

_____ 4. Add a positive and a negative number

_____ 5. Subtract two real numbers

2 Multiply and Divide Real Numbers

_____ 6. Multiply two real numbers with like signs

_____ 7. Divide two real numbers with unlike signs

_____ 8. Reciprocal of $\dfrac{a}{b}$

3 Find Powers and Square Roots of Real Numbers

_____ 9. Natural-number exponent

_____ 10. The base of -7^3

_____ 11. Principal square root of a

4 Use the Order of Operations Rule

_____ 12. Order of Operations

_____ 13. Grouping symbols

5 Evaluate Algebraic Expressions

_____ 14. To evaluate an algebraic expression

A. divide their absolute values; the quotient is negative

B. indicates how many times the base is used as a factor

C. 1) simplify within parentheses; 2) evaluate exponential expressions; 3) multiplication/division, from left to right; 4) addition/subtraction, from left to right

D. 7

E. find the numerical value, once the value of its variable(s) is known

F. add the first number to the opposite (additive inverse) of the number to be subtracted

G. the positive square root of a

H. subtract the smaller absolute value from the larger; the sign of the number with the larger absolute value is the sign of the answer

I. mathematical punctuation marks

J. $\dfrac{b}{a}$, where a and b are real numbers not equal to zero.

K. result of a subtraction problem

L. -7

M. multiply the absolute values of the two numbers; the product is positive

N. add as usual; the sum is positive

O. result of an addition problem

1.3 Operations with Real Numbers: In-Class Notes

Terms, Definitions, and Main Ideas	Examples and Notes

Use notebook paper for additional notes

Chapter 1 A Review of Basic Algebra

Important to Know

What is your homework assignment? Be sure to note it in your weekly schedule.

Section 1.3 Homework: _____ **Due**: _____

Getting to Work!

Complete your homework assignment. If you are unable to do a problem, write down the problem number and a question to help you remember what you would like to ask your instructor, your tutor, or another student.

Problem Number	Question? Where in the problem did you start to have difficulty or confusion?	Answered?

Often, you will have more questions than there is space provided here. If so, write them on notebook paper and be sure to talk to your instructor. You might ask in class or privately with the instructor.

Do You Really Know It?

Can you put into words the concepts that you learned in this section? Answer the below question from the *Writing* section in the *Study Set* in your text. Explain as if you were explaining to someone who has never taken this class before. Use notebook paper if you need more room.

Explain why the order of operations rule is necessary.

You Write the Test!

If you were writing the test for this section, what would you want a student to know? Write two test questions that you think might come from this material. Write questions of various difficulty; these questions can be original or chosen from the homework. Be sure to supply the answer also!

Write these questions at the end of this chapter under the section titled *My Practice Chapter 1 Test* and the answer to each question under the section titled *My Practice Chapter 1 Test Answers*.

Reflect on the Section

Look back at the *Pre-Class Prep* section. Did the lecture explain topics that you thought were going to be challenging or confusing? _____

- Are there topics that you still have questions on from the reading or the lecture? If so, complete the following:
    I don't understand…

- Speak to your instructor in class or during office hours about these concerns.

Reflect on Your Math Attitude

Describe your feelings toward attending class. That is, do you look forward to it, dread it, or just not really care? Why do you feel this way?

Sitting where you can see and hear the material being covered will often provide a positive experience in class.

Are You Ready?

Complete the following problems. These problems review some basic concepts that are important when simplifying expressions using properties of real numbers. *Answers are found in the Pre-Class Prep Answer Section.*

1. How do the expressions $9 \cdot x$ and $x \cdot 9$ differ?

2. How do the expressions $19 + (11 + x)$ and $(19 + 11) + x$ differ?

3. Evaluate $6(5 + 2)$ and $6 \cdot 5 + 6 \cdot 2$ and compare the results.

4. How do the terms $8x$ and $8y$ differ? What do they have in common?

 Reading Time!

While **reading Section 1.4**, identify the word or concept being defined. Choose from the following words (some may be used more than once or not at all). *Answers are found in the Pre-Class Prep Answer Section.*

simplify	division by 0	additive inverse property
term	multiplicative identity property	additive identity property
equivalent expressions	like terms	division of 0 by 0
distributive property	combining like terms	extended distributive property
division by 1	multiplication property of 0	division of a number by itself
division of 0	unlike terms	constant term
coefficient	Associative Properties of Multiplication	Commutative Properties of Multiplication
multiplicative inverse property	$ab - ac$	

1 Identify Terms, Factors, and Coefficients

1. Product or quotient of numbers and/or variables: _____

2. A term that consists of a single number: _____

3. Numerical factor of a term: _____

2 Identify and Use Properties of Real Numbers

4. Changing the order when adding or multiplying does not affect the answer: _____

5. Expressions that represent the same number: _____

6. The sum of 0 and any number is the number itself, $0 + a = a + 0 = a$: _____

7. The product of 1 and any number is the number itself, $1 \cdot a = a \cdot 1 = a$: _____

8. The product of any number and 0 is 0, $a \cdot 0 = 0 \cdot a = 0$: _____

9. For every real number a, there exists a real number $-a$ such that $a + (-a) = -a + a = 0$:

_____

10. For every nonzero real number a, there exists a real number $\dfrac{1}{a}$ such that $a \cdot \dfrac{1}{a} = \dfrac{1}{a} \cdot a = 1$:

_____

11. For any nonzero real number a, $\dfrac{a}{a} = 1$: _____

12. For any real number a, $\dfrac{a}{1} = a$: _____

13. For any nonzero real number a, $\dfrac{0}{a} = 0$: _____

14. For any nonzero real number a, $\dfrac{a}{0}$ is undefined: _____

15. $\dfrac{0}{0}$ is indeterminate: _____

3 Simplify Products

16. To write an expression in simpler form: _____

17. Two properties used to simplify certain products: _____ , _____

Chapter 1 A Review of Basic Algebra

4 Use the Distributive Property

18. If a, b, and c represent real numbers, $a(b+c) = ab + ac$: _____

19. Using the distributive property, $a(b-c) = ?$: _____

20. For any real numbers a, b, c, and so on, $a(b+c+d+e+\cdots) = ab + ac + ad + ae + \cdots$:

_____

5 Combine Like Terms

21. Terms containing exactly the same variables raised to exactly the same powers: _____

22. Terms that are not like terms: _____

23. To add or subtract the coefficients of the terms and keep the same variables with the same

exponents: _____

 Getting Ready for Class

Briefly look through the section again. Answer the following by writing the concept or just the page number from the text.

Identify concepts/procedures that you feel confident about:

Identify concepts/procedures that look confusing or challenging:

Be sure to ask your instructor further questions if you are still having difficulty with a concept.

1.4 Simplifying Algebraic Expressions Using Properties of Real Numbers: In-Class Notes

Terms, Definitions, and Main Ideas

Examples and Notes

Use notebook paper for additional notes

Chapter 1 A Review of Basic Algebra

1.4 Simplifying Algebraic Expressions Using Properties of Real Numbers: After Class

 Important to Know

What is your homework assignment? Be sure to note it in your weekly schedule.

Section 1.4 Homework: _____ **Due**: _____

Getting to Work!

Complete your homework assignment. If you are unable to do a problem, write down the problem number and a question to help you remember what you would like to ask your instructor, your tutor, or another student.

Problem Number	Question? Where in the problem did you start to have difficulty or confusion?	Answered?

Often, you will have more questions than there is space provided here. If so, write them on notebook paper and be sure to talk to your instructor. You might ask in class or privately with the instructor.

 Do You Really Know It?

Can you put into words the concepts that you learned in this section? Answer the below question from the *Writing* section in the *Study Set* in your text. Explain as if you were explaining to someone who has never taken this class before. Use notebook paper if you need more room.

Explain why the distributive property does not apply when simplifying $6(2 \cdot x)$.

You Write the Test!

     If you were writing the test for this section, what would you want a student to know? Write two test questions that you think might come from this material. Write questions of various difficulty; these questions can be original or chosen from the homework. Be sure to supply the answer also!

Write these questions at the end of this chapter under the section titled *My Practice Chapter 1 Test* **and the answer to each question under the section titled** *My Practice Chapter 1 Test Answers.*

Reflect on the Section

     Look back at the *Pre-Class Prep* section. Did the lecture explain topics that you thought were going to be challenging or confusing? _____

- Are there topics that you still have questions on from the reading or the lecture? If so, complete the following:
  I don't understand…

- Speak to your instructor in class or during office hours about these concerns.

Reflect on Your Math Attitude

     Are you pleased or disappointed with the results of your studying? _____

What would you keep the same or change about your study habits?

Studying with other students can be beneficial, as is setting aside a specific time and place.

1.5 Solving Linear Equations Using Properties of Equality: Pre-Class Prep

Are You Ready?

Complete the following problems. These review some basic skills that are needed when solving equations. *Answers are found in the Pre-Class Prep Answer Section.*

1. Simplify: $-9x - 6 + 9x$ 2. Simplify: $10a + 3 - 3$ 3. Simplify: $5m - 2(8m - 4)$

4. Multiply:
$$\frac{8}{3}\left(\frac{3}{8}x\right)$$

5. Multiply:
$$21\left(\frac{6}{7}n\right)$$

6. Multiply:
$$100 \cdot 0.27$$

Reading Time!

While **reading Section 1.5**, identify the word or concept being defined. Choose from the following words (some may be used more than once or not at all). *Answers are found in the Pre-Class Prep Answer Section.*

solution	isolated	contradiction
equation	left side	isolate the variable term on one side
right side	solution set	Subtraction Property of Equality
Addition Property of Equality	identity	clear equation of fractions and decimals
multiply by a power of 10	check the result	simplify each side of equation
equivalent equations	empty set	Division Property of Equality
$\mathbb{R}$	variable	Multiplication Property of Equality
satisfies	isolate the variables	linear equation in one variable
multiply by LCD		

1 Determine Whether a Number is a Solution

1. The set of all numbers that make the equation true: _____

2. The letter x or the unknown: _____

3. A number that makes an equation true when substituted for the variable does this for the equation: _____

4. A number that makes an equation true when substituted for the variable: _____

5. A statement indicating that two expressions are equal: _____

6. For the equation $x + 4 = 12$, the side that 12 is on: _____

2 Use Properties of Equality to Solve Equations

7. $ax + b = c$, where a, b, and c are real numbers, and $a \neq 0$: _____

8. Equations with the same solutions: _____

9. For any real numbers a, b, and c, if $a = b$, then $a + c = b + c$: _____

10. For any real numbers a, b, and c, if $a = b$, then $a - c = b - c$: _____

11. For any real numbers a, b, and c, where c is not 0, if $a = b$, then $ca = cb$: _____

12. For any real numbers a, b, and c, where c is not 0, if $a = b$, then $\dfrac{a}{c} = \dfrac{b}{c}$: _____

13. When a variable is alone or by itself on one side of the equation, the variable is..: _____

3 Simplify Expressions to Solve Equations

Strategy for Solving Linear Equations in One Variable:

14. Step 1: _____

15. Step 2: _____

16. Step 3: _____

17. Step 4: _____

18. Step 5: _____

4 Clear Equations of Fractions and Decimals

19. To clear an equation of fractions: _____

20. To clear an equation of decimals: _____

5 Identify Identities and Contradictions

21. An equation made true by any permissible replacement value for the variable: _____

22. An equation made false for all replacement values for the variable: _____

23. The solution set of a contradiction: _____

24. The solution set of an identity: _____

✓ Getting Ready for Class

Briefly look through the section again. Answer the following by writing the concept or just the page number from the text.

Identify concepts/procedures that you feel confident about:

Identify concepts/procedures that look confusing or challenging:

Be sure to ask your instructor further questions if you are still having difficulty with a concept.

1.5 Solving Linear Equations Using Properties of Equality: In-Class Notes

Terms, Definitions, and Main Ideas	Examples and Notes

Use notebook paper for additional notes

1.5 Solving Linear Equations Using Properties of Equality: After Class

Important to Know

What is your homework assignment? Be sure to note it in your weekly schedule.

Section 1.5 Homework: _____ **Due:** _____

Getting to Work!

Complete your homework assignment. If you are unable to do a problem, write down the problem number and a question to help you remember what you would like to ask your instructor, your tutor, or another student.

Problem Number	Question? Where in the problem did you start to have difficulty or confusion?	Answered?

Often, you will have more questions than there is space provided here. If so, write them on notebook paper and be sure to talk to your instructor. You might ask in class or privately with the instructor.

Do You Really Know It?

Can you put into words the concepts that you learned in this section? Answer the below question from the *Writing* section in the *Study Set* in your text. Explain as if you were explaining to someone who has never taken this class before. Use notebook paper if you need more room.

When solving a linear equation in one variable, the objective is to isolate the variable on one side of the equation. What does that mean?

You Write the Test!

If you were writing the test for this section, what would you want a student to know? Write test questions that you think might come from this material. Write questions of various difficulty; these questions can be original or chosen from the homework. Be sure to supply the answer also!

Write these questions at the end of this chapter under the section titled *My Practice Chapter 1 Test* and the answer to each question under the section titled *My Practice Chapter 1 Test Answers*.

Reflect on the Section

Look back at the *Pre-Class Prep* section. Did the lecture explain topics that you thought were going to be challenging or confusing? _____

- Are there topics that you still have questions on from the reading or the lecture? If so, complete the following:
    I don't understand…

- Speak to your instructor in class or during office hours about these concerns.

Reflect on Your Math Attitude

Have you built your support system? In this chapter's *Study Skills Workshop*, the *Support System Worksheet* and the *Peer Contact Information Guide* were highlighted. Complete these if you have not already done so.

How has having this information influenced your attitude about your success with this course?

Developing a working relationship with your instructor and other students can be very helpful.

1.6 Solving Formulas; Geometry: Pre-Class Prep

Are You Ready?

Complete the following problems. These problems review some basic skills that are needed when working with formulas. *Answers are found in the Pre-Class Prep Answer Section.*

Determine which concept (perimeter, area, volume, or circumference) should be used to find each of the following.

1. The amount of storage in a refrigerator

2. The distance the tip of an airplane propeller travels

3. The distance around a dance floor

4. The amount of floor space to carpet

Reading Time!

(on the next page)

Getting Ready for Class

Briefly look through the section again. Answer the following by writing the concept or just the page number from the text.

Identify concepts/procedures that you feel confident about:

Identify concepts/procedures that look confusing or challenging:

Be sure to ask your instructor further questions if you are still having difficulty with a concept.

Reading Time!

While **reading Section 1.6**, match the word or concept to its definition or description. Not all choices are used. *Answers are found in the Pre-Class Prep Answer Section.*

1 Use Formulas from Geometry

_____ 1. Perimeter

_____ 2. Perimeter formulas

_____ 3. Area

_____ 4. Area formulas

_____ 5. Diameter of a circle

_____ 6. Radius of a circle

_____ 7. Circumference

_____ 8. Circle formulas

_____ 9. Volume

_____ 10. Volume formulas

2 Solve for a Specified Variable

_____ 11. Solve a formula for a specified variable

_____ 12. "solved for w"

3 Solve Application Problems Using Formulas

_____ 13. $F = \frac{9}{5}C + 32$

_____ 14. Solve $F = \frac{9}{5}C + 32$ for C

A. chord that passes through the center

B. $D = 2r$ (diameter); $r = \frac{1}{2}D$ (radius); $C = 2\pi r$ and $C = \pi D$ (circumference); $A = \pi r^2$ (area)

C. amount of space that the geometric solid encloses

D. $C = \frac{5}{9}(F - 32)$

E. isolate the variable on one side of the equation, with all other variables and constants on the opposite side

F. $P = 2l + 2w$ (rectangle); $P = 4s$ (square); $P = a + b + c$ (triangle)

G. w is alone on one side of the equation and the other side does not contain w

H. $A = \frac{1}{2}(bh)$ (triangle); $A = lw$ (rectangle); $A = \frac{1}{2}h(b_1 + b_2)$ (trapezoid); $A = s^2$ (square)

I. segment drawn from the center of a circle to a point on the circle

J. $V = lwh$ (rectangular solid); $V = s^3$ (cube); $V = \frac{4}{3}\pi r^3$ (sphere); $V = \pi r^2 h$ (cylinder)

K. amount of surface that the geometric figure encloses

L. distance around a circle

M. formula that relates Fahrenheit to Celsius temperature

N. sum of the lengths of the sides of a geometric figure

1.6 Solving Formulas; Geometry: In-Class Notes

Terms, Definitions, and Main Ideas	Examples and Notes

Use notebook paper for additional notes

1.6 Solving Formulas; Geometry: After Class

Important to Know

What is your homework assignment? Be sure to note it in your weekly schedule.

Section 1.6 Homework: _____ **Due:** _____

Getting to Work!

Complete your homework assignment. If you are unable to do a problem, write down the problem number and a question to help you remember what you would like to ask your instructor, your tutor, or another student.

Problem Number	Question? Where in the problem did you start to have difficulty or confusion?	Answered?

Often, you will have more questions than there is space provided here. If so, write them on notebook paper and be sure to talk to your instructor. You might ask in class or privately with the instructor.

Do You Really Know It?

Can you put into words the concepts that you learned in this section? Answer the below question from the *Writing* section in the *Study Set* in your text. Explain as if you were explaining to someone who has never taken this class before. Use notebook paper if you need more room.

Explain the error made below.

$$T = \frac{adx + y}{x}$$

Chapter 1 A Review of Basic Algebra

You Write the Test!

If you were writing the test for this section, what would you want a student to know? Write two test questions that you think might come from this material. Write questions of various difficulty; these questions can be original or chosen from the homework. Be sure to supply the answer also!

Write these questions at the end of this chapter under the section titled *My Practice Chapter 1 Test* and the answer to each question under the section titled *My Practice Chapter 1 Test Answers*.

Reflect on the Section

Look back at the *Pre-Class Prep* section. Did the lecture explain topics that you thought were going to be challenging or confusing? _____

- Are there topics that you still have questions on from the reading or the lecture? If so, complete the following:
    I don't understand…

- Speak to your instructor in class or during office hours about these concerns.

Reflect on Your Math Attitude

You have been tracking weekly how much time you spend studying *(Target Goal)*. Are you pleased with the amount of time that you have spent studying this section?

Do you feel you are spending too much time or too little time on this section? What changes could you make?

Seek advice on effective studying from your instructor, tutor, or other campus resources.

1.7 Using Equations to Solve Problems: Pre-Class Prep

✓ Are You Ready?

Complete the following problems. These problems review some basic skills that are needed to solve the application problems in this section. *Answers are found in the Pre-Class Prep Answer Section.*

1. Translate to symbols: *8 less than twice a number x*

2. If $x = 15$, find $6x + 3$.

3. If mouse pads cost $5.95 each, what is the cost of 10 of them?

4. What is the formula for the perimeter of a rectangle?

5. What is the sum of the measures of the angles of a triangle?

6. Solve: $2x + 10 + 53x + 9 + x = 1,251$

✓ Reading Time!

While **reading Section 1.7**, fill in the blanks choosing from the following words (some may be used more than once or not at all). *Answers are found in the Pre-Class Prep Answer Section.*

mathematical	facts	vertex	solve	analyze	complement
supplementary	base angles	assign	equilateral	isosceles	check
unknown	sum	equation	scalene	original	conclusion
straight angle	acute angle	table	number · value	right angle	obtuse angle
supplement	right	180°	90°	0°	

Chapter 1 A Review of Basic Algebra

1 Apply the Steps of a Problem-Solving Strategy

1. _____ the problem by reading it carefully to understand the given _____.

2. _____ a variable to represent an _____ value in the problem.

3. Form an _____ by translating the words of the problem into _____ symbols.

4. _____ the equation formed in step 3 of the strategy for problem solving.

5. State the _____ clearly.

6. _____ the result using the _____ wording of the problem.

2 Solve Number-Value Problems

7. The total value of an item is _____.

8. It is helpful to use a _____ to organize facts of an application problem.

3 Solve Geometry Problems

9. A _____ is an angle with measure $90°$, while a _____ has a measure of $180°$.

10. An _____ is an angle whose measure is greater than $0°$ and less than $90°$; an _____ has measure greater than $90°$ and less than $180°$.

11. Complementary angles sum to _____, and each angle is called the _____ of the other.

12. _____ angles sum to $180°$, and each angle is called the _____ of the other.

13. An _____ triangle has two sides of equal length that meet to form the _____ angle.

14. A triangle with one right angle is called a _____ triangle; a triangle with three equal sides and three equal angles is called an _____ triangle.

15. Angles opposite equal sides of a triangle are called _____; these angles are equal.

16. The _____ of the measures of the angles of any triangle is $180°$.

Getting Ready for Class

Briefly look through the section again. Answer the following by writing the concept or just the page number from the text.

Identify concepts/procedures that you feel confident about:

Identify concepts/procedures that look confusing or challenging:

Be sure to ask your instructor further questions if you are still having difficulty with a concept.

Chapter 1 A Review of Basic Algebra

Terms, Definitions, and Main Ideas	Examples and Notes

Use notebook paper for additional notes

1.7 Using Equations to Solve Problems: After Class

Important to Know

What is your homework assignment? Be sure to note it in your weekly schedule.

Section 1.7 Homework: _____ **Due:** _____

Getting to Work!

Complete your homework assignment. If you are unable to do a problem, write down the problem number and a question to help you remember what you would like to ask your instructor, your tutor, or another student.

Problem Number	Question? Where in the problem did you start to have difficulty or confusion?	Answered?

Often, you will have more questions than there is space provided here. If so, write them on notebook paper and be sure to talk to your instructor. You might ask in class or privately with the instructor.

Do You Really Know It?

Can you put into words the concepts that you learned in this section? Answer the below question from the *Writing* section in the *Study Set* in your text. Explain as if you were explaining to someone who has never taken this class before. Use notebook paper if you need more room.

Briefly explain what should be accomplished in each of the steps (analyze, assign, form, solve, state, and check) of the problem-solving strategy used in this section.

You Write the Test!

If you were writing the test for this section, what would you want a student to know? Write two test questions that you think might come from this material. Write questions of various difficulty; these questions can be original or chosen from the homework. Be sure to supply the answer also!

Write these questions at the end of this chapter under the section titled *My Practice Chapter 1 Test* and the answer to each question under the section titled *My Practice Chapter 1 Test Answers*.

 Reflect on the Section

Look back at the *Pre-Class Prep* section. Did the lecture explain topics that you thought were going to be challenging or confusing? _____

- Are there topics that you still have questions on from the reading or the lecture? If so, complete the following:
    I don't understand…

- Speak to your instructor in class or during office hours about these concerns.

Reflect on Your Math Attitude

How do you feel about your homework? That is, do you feel that you are in control or that the homework is in control of you?

Describe what has or has not worked for you with respect to your homework. Would you make any changes?

Evaluate your work in this course and be willing to try new approaches to completing your homework.

1.8 More about Problem Solving: Pre-Class Prep

 Are You Ready?

Complete the following problems. These problems review some basic skills that are needed in this section. *Answers are found in the Pre-Class Prep Answer Section.*

1. a. Write 27.8% as a decimal.   b. Write 0.8356 as a percent.

2. Find the amount of interest earned by $15,000 invested at a 2% annual simple interest rate for 1 year.

3. At 55 miles per hour, how far will a car travel in 4 hours?

4. A pharmacist has 2 liters of a 30% alcohol solution. How many liters of pure alcohol are in the solution?

5. At $5.55 per pound, what is the value of 6 pounds of roast beef?

6. A couple invested $x of the $60,000 lottery winnings in bonds. How much do they have left to invest in stocks?

 Reading Time!

While **reading Section 1.8**, identify the word or concept being defined. Choose from the following words (some may be used more than once or not at all). *Answers are found in the Pre-Class Prep Answer Section.*

concentrations	original	total value = amount · price
$I = Prt$	regular price	principal
$d = rt$	dry mixture	rate
changed	uniform motion	amount of solution · strength of the solution
percent	markdown	percent sentence

1 Solve Percent Problems

1. The word _____ means parts per one hundred.

2. To solve applied percent problems, one method is to use the given facts to write a _____.

3. The amount of the reduction in the price of an item is called a _____ .

4. The sale price is equal to the _____ minus the _____. The markdown is expressed as a _____ of the regular price.

5. Percents are often used to describe how a quantity has _____.

6. The percent of increase (or decrease) is a percent of the _____ amount.

2 Solve Investment Problems

7. The amount of simple interest I an investment earns is given by the formula _____.

8. The amount invested is called the _____.

3 Solve Uniform Motion Problems

9. A _____ problem involves an object traveling at a constant rate for a specified period of time over a certain distance.

10. The formula _____ is used to solve uniform motion problems.

4 Solve Mixture Problems

11. A _____ is created from two differently priced ingredients (or components), where the total value is given by the formula _____.

12. A liquid mixture of a desired strength is to be made from two solutions with different strengths, or _____.

13. The amount of a pure substance (such as butterfat) is _____.

Getting Ready for Class

Briefly look through the section again. Answer the following by writing the concept or just the page number from the text.

Identify concepts/procedures that you feel confident about:

Identify concepts/procedures that look confusing or challenging:

Be sure to ask your instructor further questions if you are still having difficulty with a concept.

1.8 More about Problem Solving: In-Class Notes

Terms, Definitions, and Main Ideas	Examples and Notes

Use notebook paper for additional notes

1.8 More about Problem Solving: After Class

Important to Know

What is your homework assignment? Be sure to note it in your weekly schedule.

Section 1.8 Homework: _____ **Due:** _____

Getting to Work!

Complete your homework assignment. If you are unable to do a problem, write down the problem number and a question to help you remember what you would like to ask your instructor, your tutor, or another student.

Problem Number	Question? Where in the problem did you start to have difficulty or confusion?	Answered?

Often, you will have more questions than there is space provided here. If so, write them on notebook paper and be sure to talk to your instructor. You might ask in class or privately with the instructor.

Do You Really Know It?

Can you put into words the concepts that you learned in this section? Answer the below question from the *Writing* section in the *Study Set* in your text. Explain as if you were explaining to someone who has never taken this class before. Use notebook paper if you need more room.

If a mixture is to be made from solutions with concentrations of 12% and 30%, can the mixture have a concentration less than 12%. Can the mixture have a concentration greater than 30%? Explain.

You Write the Test!

If you were writing the test for this section, what would you want a student to know? Write two test questions that you think might come from this material. Write questions of various difficulty; these questions can be original or chosen from the homework. Be sure to supply the answer also!

Write these questions at the end of this chapter under the section titled *My Practice Chapter 1 Test* and the answer to each question under the section titled *My Practice Chapter 1 Test Answers*.

Reflect on the Section

Look back at the *Pre-Class Prep* section. Did the lecture explain topics that you thought were going to be challenging or confusing? _____

- Are there topics that you still have questions on from the reading or the lecture? If so, complete the following:
    I don't understand…

- Speak to your instructor in class or during office hours about these concerns.

Reflect on Your Math Attitude

This is the last section of this chapter. It is quite possible that you will be having an exam or quiz over this material. Describe your thoughts and concerns about this.

Speak with your instructor if you have concerns. Read the *Chapter 1 Test Skills Assessment* for study suggestions. Now is a good time to begin taking your *My Practice Chapter 1 Test*.

Chapter 1 Activities

 Your instructor may assign these activities to you to complete in class, or you may complete them on your own to solidify your understanding of chapter topics. The activities begin on the next page.

❖ **Student Activity:** *Venn Diagram of the Real Numbers*
> Place numbers in the smallest set (natural numbers, whole numbers, integers, rational numbers, irrational numbers, or real numbers) in which it belongs.

❖ **Student Activity:** *Pick Your Property*
> Match up equations with the proper property of the real numbers.

❖ **Student Activity:** *What's Wrong with Division by Zero?*
> Investigate why division by zero is undefined.

Student Activity
Venn Diagram of the Real Numbers

Directions: Place each number below in the *smallest* set in which it belongs. For example, −1 is a real number, a rational number, and an integer, so we place it in the "Integers" box, but not inside the whole numbers or natural numbers.

$$8 \qquad \frac{7}{3} \qquad -1.3 \qquad \pi \qquad 2.175 \qquad 0 \qquad -7 \qquad \sqrt{2} \qquad 1000 \qquad 0.00005$$

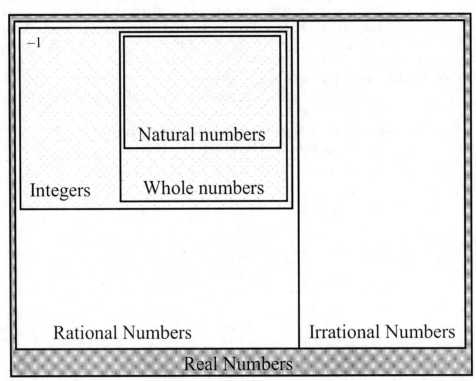

1. Given all possible real numbers, name at least one number that is a whole number, but not a natural number: _____

2. Can a number be both rational and irrational? _____ If yes, name one: _____

3. Can a number be both rational and an integer? _____ If yes, name one: _____

4. Given all possible real numbers, name at least one number that is an integer, but not a whole number: _____

Side note: *Just for the record, this diagram in no way conveys the actual size of the sets. In mathematics, the number of elements that belong to a set is called the **cardinality** of the set. Technically (and with a lot more mathematics classes behind you) it can be proven that the cardinality of the irrational numbers (uncountable infinity) is actually larger than the cardinality of the rational numbers (countable infinity). Another interesting fact is that the cardinality (size) of the rational numbers, integers, whole numbers, and natural numbers are all equal. This type of mathematics is studied in a course called Real Analysis (that comes after the Calculus sequence).*

Student Activity
Pick Your Property

Match-up: Match each of the equations in the squares of the table below with the proper property of the real numbers.

A Associative Property of Addition

B Associative Property of Multiplication

C Commutative Property of Addition

D Commutative Property of Multiplication

E Inverse Property of Addition

F Inverse Property of Multiplication

G Identity Property of Addition

H Identity Property of Multiplication

I Multiplication Property of Zero

$0 + (-2) = -2$	$77(-2) = (-2)77$	$\left(\frac{1}{4} + \frac{2}{3}\right) + \frac{1}{3} = \frac{1}{4} + \left(\frac{2}{3} + \frac{1}{3}\right)$
$2\left(\frac{3}{100}\right)\left(\frac{100}{3}\right) = 2(1)$	$8(-9 + 9) = 8(0)$	$\left(\frac{1}{4} + \frac{2}{3}\right) + \frac{1}{3} = \left(\frac{2}{3} + \frac{1}{4}\right) + \frac{1}{3}$
$\frac{4}{5}\left(\frac{5}{4}\right) = \frac{5}{4}\left(\frac{4}{5}\right)$	$5 + (6 + (-6)) = (5 + 6) + (-6)$	$\left(2 \cdot \frac{3}{100}\right)\left(\frac{100}{3}\right) = 2 \cdot \left(\frac{3}{100} \cdot \frac{100}{3}\right)$
$5\left(\frac{1}{5} \cdot 1\right) = 5\left(\frac{1}{5}\right)$	$-5 + (-8 + 2) = (-8 + 2) + (-5)$	$5 + (6 + (-6)) = 5 + 0$
$\frac{4}{5}\left(\frac{5}{4}\right) = 1$	$5 + 9(0) = 5 + 0$	$24 + 6(1) = 24 + 6$

Student Workbook Activities, M. Andersen

Chapter 1 A Review of Basic Algebra

Student Activity
What's Wrong with Division by Zero?

Let's spend some time investigating why division by zero is undefined. You will need a calculator for this activity.

1. First, let's see what your calculator thinks. Try the following division problems on your calculator and write down the results:

$$0 \div 5 \qquad 5 \div 0 \qquad \frac{12}{0} \qquad \frac{0}{12}$$

2. Even your calculator will reject the idea of division by zero, so let's try dividing by numbers *close* to zero. Looking at the number line below, name and label some numbers that are really close to zero (on both sides of zero).

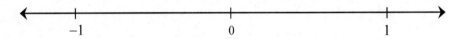

3. Find the following quotients using your calculator.

$$5 \div 0.1 \qquad 5 \div 0.01 \qquad 5 \div 0.001 \qquad 5 \div 0.0001$$

4. What is happening to the quotients in #3 as the divisor gets closer to zero?

5. Find the following quotients using your calculator.

$$5 \div (-0.1) \qquad 5 \div (-0.01) \qquad 5 \div (-0.001) \qquad 5 \div (-0.0001)$$

6. What is happening to the quotients in problem **5** as the divisor gets closer to zero?

7. Using the results from problems **4** and **6**, why do you think that division by zero is undefined?

Chapter 1 Test Skills Assessment

Pre-Test Preparation Work:

1. Re-read the objectives from each section.
2. Review the *Reading Time!* activity for each section.
3. Go over all your classroom notes, if something in your notes doesn't make sense to you, make a note and ask your teacher or a classmate.
4. Make additional notations to your work if your teacher states specific concepts to study in preparation for the chapter test.
5. Attend any study sessions held by your teacher or teaching assistant.
6. Practice additional problems.
7. Go over any missed problems in your homework sets.
8. Talk out concepts with your peers in small group study sessions.

List other preparations that you have found beneficial in preparing for a math test.

Additional Practice Suggestions:

1. Use your book's review problems at the back of the chapter as a practice test.
2. Take *My Practice Chapter Test* and the text's *Chapter Test*. Time yourself and do not use your notebook or textbook.
3. Pace yourself as you work through these problems.
4. Read each question carefully, playing close attention to the instructions.
5. Check your work using the answers provided.
6. Rework any missed problems. Do not just "look them over" but actually rework the problem without looking at text or notes.

Chapter 1 A Review of Basic Algebra

My Practice Chapter 1 Test

For each section, you had the opportunity to create two test questions under the section *You Write the Test!* Write each of those questions here. Include your answers under the heading *My Practice Chapter 1 Test Answers.* **Take the test without notes or your textbook.** If you do not get a question correct, review the text and/or your notes then take the test again. For further review, do the *Chapter 1 Test* in the text.

Section 1.1

1.

2.

Section 1.2

3.

4.

Section 1.3

5.

6.

Section 1.4

7.

8.

Section 1.5

9.

10.

Section 1.6

11.

12.

Section 1.7

13.

14.

Section 1.8

15.

16.

Chapter 1 A Review of Basic Algebra

Section 1.1

1.

2.

Section 1.2

3.

4.

Section 1.3

5.

6.

Section 1.4

7.

8.

Section 1.5

9.

10.

Section 1.6

11.

12.

Section 1.7

13.

14.

Section 1.8

15.

16.

Chapter 2 Graphs, Equations of Lines, and Functions

Read the *Study Skills Workshop* found at the beginning of Chapter 2 in your textbook. **Complete** the activities below for this chapter's *Study Skills Workshop*.

Preparing to Learn: Learn your individual learning style to help you to be successful in math!

Discover Your Learning Style

Your individual learning style is how you learn concepts the best. For instance, if you are an auditory learner, you would prefer to hear instructions rather than to read them. So, the question is, what type of learner are YOU?

✓ Take *A Learning Style Survey for College* found at

http://www.metamath.com/multiple/multiple_choice_questions.html

Complete the following once you have completed the survey.

My primary learning style is _____.

Some learning strategies for me would be: *(write on another page if you need more room)*

1. _____

2. _____

3. _____

Get the Most Out of the Textbook

✓ Take the *Textbook Tour* found online at www.cengage.com/math/tussy.

Take Good Notes

Good notes help when you are trying to remember important concepts.

✓ Read your notes from your last lecture. Did your notes make sense? _____

Rewrite a set of your class notes to make them more readable and to clarify the concepts and examples covered. Fill in any information you didn't have time to copy down in class and complete any phrases or sentence fragments.

✓ Complete the *Pre-Class Prep* sections in this notebook. This will help you to take better notes as you will know what to expect.

On the Internet, additional note-taking hints can be found by searching "how to take lecture notes" or "how to take effective class notes."

2.1 Graphs: Pre-Class Prep

 Are You Ready?

Complete the following problems. These review some basic skills that are needed when graphing ordered pairs. *All answers are found in the Pre-Class Prep Answer Section.*

1. Graph each number in the set $\left\{\dfrac{5}{3}, -2, 0, 3, -3.75\right\}$ on a number line.

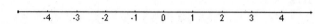

2. a. What number is 6 units to the right of 0 on a number line?

   b. What number is 4.5 units to the left of 0 on a number line?

3. List the first four Roman numerals.

4. Write $\dfrac{7}{2}$ and $-\dfrac{16}{5}$ in mixed-number form.

 Reading Time!

While **reading Section 2.1**, identify the word or concept being defined. Choose from the following words (some may be used more than once or not at all). *All answers are found in the Pre-Class Prep Answer Section.*

graphing/plotting	y-axis	(4, –6)
$\left(\dfrac{x_1 - x_2}{2}, \dfrac{y_1 - y_2}{2}\right)$	y-coordinate	quadrants
paired data	x-axis	(4, 6)
origin	$\left(\dfrac{x_1 + x_2}{2}, \dfrac{y_1 + y_2}{2}\right)$	ordered pair
coordinate plane	(–4, 6)	coordinates
midpoint	x-coordinate	(–4, –6)
graph	endpoints	step graph

1 Plot Ordered Pairs and Determine the Coordinates of a Point

1. The four regions in a coordinate plane: _____

2. The horizontal line in a rectangular coordinate system: _____

3. Formed by two perpendicular number lines: _____

4. The second number y in an ordered pair: _____

5. The vertical line in a rectangular coordinate system: _____

6. Identifies each point in the coordinate plane: _____

7. The point where the axes intersect: _____

8. The x-coordinate and the y-coordinate together are called: the _____ of the point

9. Process of locating a point in the coordinate plane: _____

10. Starting at the origin, the ordered pair for the point that is 4 units to the left on the x-axis and 6 units up on the y-axis: _____

2 Graph Paired Data

11. Data that can be represented as an ordered pair: _____

3 Read Graphs

12. A popular way to present information: _____

13. A graph consisting of horizontal line segments: _____

4 Find the Midpoint of a Line Segment

14. A point that lies midway between two other points: _____

15. If M is the midpoint between P and Q, P and Q are called the: _____

16. The Midpoint Formula: _____

✓ Getting Ready for Class

Briefly look through the section again. Answer the following by writing the concept or just the page number from the text.

Identify concepts/procedures that you feel confident about:

Identify concepts/procedures that look confusing or challenging:

Be sure to ask your instructor further questions if you are still having difficulty with a concept.

2.1 Graphs: In-Class Notes

Terms, Definitions, and Main Ideas

Examples and Notes

Use notebook paper for additional notes

2.1 Graphs: After Class

Important to Know

What is your homework assignment? Be sure to note it in your weekly schedule.

Section 2.1 Homework: _____ **Due**: _____

Getting to Work!

Complete your homework assignment. If you are unable to do a problem, write down the problem number and a question to help you remember what you would like to ask your instructor, your tutor, or another student.

Problem Number	Question? Where in the problem did you start to have difficulty or confusion?	Answered?

Often, you will have more questions than there is space provided here. If so, write them on notebook paper and be sure to talk to your instructor. You might ask in class or privately with the instructor.

Do You Really Know It?

Can you put into words the concepts that you learned in this section? Answer the below question from the *Writing* section in the *Study Set* in your text. Explain as if you were explaining to someone who has never taken this class before. Use notebook paper if you need more room.

Explain why the coordinates of the origin are (0,0).

You Write the Test!

If you were writing the test for this section, what would you want a student to know? Write test questions that you think might come from this material. Write questions of various difficulty; these questions can be original or chosen from the homework. Be sure to supply the answer also!

Write these questions at the end of this chapter under the section titled *My Practice Chapter 2 Test* and the answer to each question under the section titled *My Practice Chapter 2 Test Answers*.

Reflect on the Section

Look back at the *Pre-Class Prep* section. Did the lecture explain topics that you thought were going to be challenging or confusing? _____

- Are there topics that you still have questions on from the reading or the lecture? If so, complete the following:
  I don't understand…

- Speak to your instructor in class or during office hours about these concerns.

Reflect on Your Math Attitude

According to the *Learning Style Survey*, what is your learning style?

How has knowing how you learn affected your attitude toward math?

It is good to know that not everyone learns the same, so it is just fine if your style is different from another student's learning style.

2.2 Graphing Linear Equations in Two Variables: Pre-Class Prep

✓ Are You Ready?

Complete the following problems. These review some basic skills that are needed when graphing linear equations. *Answers are found in the Pre-Class Prep Answer Section.*

1. Is $4(2)+-3(-1)=10$ a true or false statement?

2. Solve: $5(3)+3y=6$

3. Graph the points $(2,0),(-4,0),(1,0),$ and $(0,-3)$.

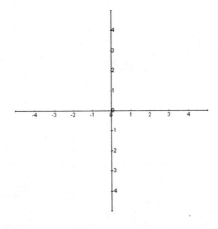

4. What is the x-coordinate of any point that lies on the y-axis?

5. Solve: $8(0)-5y=10$

6. Draw a vertical line. Draw a horizontal line.

Reading Time!

While **reading Section 2.2**, fill in the blanks choosing from the following words (some may be used more than once or not at all). *All answers are found in the Pre-Class Prep Answer Section.*

solution	plotting/plot	two	equation	declining
only one	all	linear models	ordered pair	straight-line depreciation
y	real-life	satisfies	standard	infinitely many
graph	$A + B = C$	zero (0)	x	$Ax + By = C$
variables	corresponding	linear	three	select a number
$(0, b)$	straight line	intercept method	y-axis	is / is not
horizontal	$(a, 0)$	x-axis	vertical	

1 Determine Whether an Ordered Pair is a Solution of an Equation

1. A _____ of an equation in two variables is a(n) _____ of numbers that makes the equation true.

2. The equation $y = x - 1$ is an equation in _____ variables.

3. If an ordered pair makes the equation true, then that ordered pair _____ the equation.

2 Find a Solution of an Equation in Two Variables

4. To find a _____ of an equation in two variables, _____ for one of the variables and find the _____ value of the other variable.

5. Since an equation in two variables has _____ solutions, we draw a mathematical picture of the solutions, called the _____ of the equation.

3 Graph Linear Equations by Plotting Points

6. The graph of an equation is a drawing that represents _____ of its solutions.

7. To graph a linear equation, find _____ ordered pairs that are solutions of the equation, _____ the solutions on a rectangular coordinate system, and draw a _____ passing through the points.

8. The graph of an equation in two variables is a _____ and so the equation is said to be _____.

9. A _____ equation in two variables can be written in the form _____, where A, B, and C are real numbers and A and B are not both _____. This form is called _____ form.

4 Graph Linear Equations by Finding Intercepts

10. The _____ of graphing a line is plotting the x- and y-intercepts of a graph and drawing a line through them.

11. The y-intercept of a line is the point _____, where the line intersects the _____. To find b, substitute _____ for x in the equation of the line and solve for _____.

12. The x-intercept of a line is the point _____, where the line intersects the _____. To find a, substitute _____ for y in the equation of the line and solve for _____.

5 Graph Horizontal and Vertical Lines

13. The graph of $y = 0$ is the _____.

14. The equation of a _____ line is $y = b$.

15. The equation of a _____ line is $x = a$.

16. The equation $x = 4$ _____ a linear equation in two variables.

5 Use Linear Models to Solve Applied Problems

17. To model _____ situations, linear equations, or _____, are often written using _____ other than x and y.

18. The method called _____ can be used to find the _____ value of aging equipment. The straight-line depreciation _____ determines the value.

Getting Ready for Class

Briefly look through the section again. Answer the following by writing the concept or just the page number from the text.

Identify concepts/procedures that you feel confident about:

Identify concepts/procedures that look confusing or challenging:

Be sure to ask your instructor further questions if you are still having difficulty with a concept.

2.2 Graphing Linear Equations in Two Variables: In-Class Notes

Terms, Definitions, and Main Ideas	Examples and Notes

Use notebook paper for additional notes

2.2 Graphing Linear Equations in Two Variables: After Class

 Important to Know

What is your homework assignment? Be sure to note it in your weekly schedule.

Section 2.2 Homework: _____ **Due**: _____

 Getting to Work!

Complete your homework assignment. If you are unable to do a problem, write down the problem number and a question to help you remember what you would like to ask your instructor, your tutor, or another student.

Problem Number	Question? Where in the problem did you start to have difficulty or confusion?	Answered?

Often, you will have more questions than there is space provided here. If so, write them on notebook paper and be sure to talk to your instructor. You might ask in class or privately with the instructor.

 Do You Really Know It?

Can you put into words the concepts that you learned in this section? Answer the below question from the *Writing* section in the *Study Set* in your text. Explain as if you were explaining to someone who has never taken this class before. Use notebook paper if you need more room.

Explain how to graph a line using the intercept method.

You Write the Test!

If you were writing the test for this section, what would you want a student to know? Write two test questions that you think might come from this material. Write questions of various difficulty; these questions can be original or chosen from the homework. Be sure to supply the answer also!

Write these questions at the end of this chapter under the section titled *My Practice Chapter 2 Test* and the answer to each question under the section titled *My Practice Chapter 2 Test Answers*.

Reflect on the Section

Look back at the *Pre-Class Prep* section. Did the lecture explain topics that you thought were going to be challenging or confusing? _____

- Are there topics that you still have questions on from the reading or the lecture? If so, complete the following:
    I don't understand...

- Speak to your instructor in class or during office hours about these concerns.

Reflect on Your Math Attitude

Think about the learning strategies that you identified once you knew your learning style. Have you implemented any of them? If so, which one. If not, what would you like to implement?

How has utilizing a learning strategy that fits you affected your attitude toward your learning?

2.3 Rate of Change and the Slope of a Line: Pre-Class Prep

Are You Ready?

Complete the following problems. These problems review some basic skills that are needed to find the slope of a line. *All answers are found in the Pre-Class Prep Answer Section.*

1. Evaluate: $\dfrac{4-1}{8-3}$

2. Evaluate: $\dfrac{-10-1}{-4-(-4)}$

3. Simplify: $\dfrac{24}{28}$

4. Fill in the blanks. _____ lines do not intersect. _____ lines intersect to form right angles.

Reading Time!

(on the next page)

Getting Ready for Class

Briefly look through the section again. Answer the following by writing the concept or just the page number from the text.

Identify concepts/procedures that you feel confident about:

Identify concepts/procedures that look confusing or challenging:

Be sure to ask your instructor further questions if you are still having difficulty with a concept.

Reading Time!

While **reading Section 2.3**, match the word or concept to its definition or description. Not all choices are used. *All answers are found in the Pre-Class Prep Answer Section.*

1 Calculate an Average Rate of Change

_____ 1. Rate of change

_____ 2. Ratio

_____ 3. Rate

2 Find the Slope of a Line from Its Graph

_____ 4. Slope

_____ 5. Rise

_____ 6. Slope triangle

3 Find the Slope of a Line Given Two Points

_____ 7. Change in x

_____ 8. Subscript notation

_____ 9. Slope of a line

_____ 10. Symbol Δ

4 Find the Slope of Horizontal and Vertical Lines

_____ 11. Slope of vertical line

_____ 12. Slope of horizontal line

5 Solve Applications of Slope

_____ 13. Grade

6 Determine Whether Lines Are Parallel or Perpendicular Using Slope

_____ 14. Slopes of nonvertical perpendicular lines

_____ 15. Perpendicular lines

_____ 16. Slopes of nonvertical parallel lines

_____ 17. Negative reciprocals

A. comparison of 2 numbers using a quotient

B. Greek letter *delta* used to represent change as in Δy and Δx.

C. 0

D. comparison of the change in one quantity with respect to another

E. slope of a incline expressed as a percent

F. vertical change between two points

G. used to distinguish between the coordinates of two points

H. negative reciprocals

I. are the same

J. $y_2 - y_1$

K. undefined slope

L. $m = \dfrac{\text{change in } y}{\text{change in } x} = \dfrac{y_2 - y_1}{x_2 - x_1}$, if $x_2 \neq x_1$

M. right triangle created showing vertical and horizontal change between two points

N. $x_2 - x_1$

O. two numbers whose product is –1

P. lines that intersect to form four right angles

Q. ratio that compares the vertical change with the horizontal change moving from one point along the line to another

R. ratios that are used to compare quantities with different units.

2.3 Rate of Change and the Slope of a Line: In-Class Notes

Terms, Definitions, and Main Ideas	Examples and Notes

Use notebook paper for additional notes

Chapter 2 Graphs, Equations of Lines, and Functions

2.3 Rate of Change and the Slope of a Line: After Class

Important to Know

What is your homework assignment? Be sure to note it in your weekly schedule.

Section 2.3 Homework: _____ **Due**: _____

Getting to Work!

Complete your homework assignment. If you are unable to do a problem, write down the problem number and a question to help you remember what you would like to ask your instructor, your tutor, or another student.

Problem Number	Question? Where in the problem did you start to have difficulty or confusion?	Answered?

Often, you will have more questions than there is space provided here. If so, write them on notebook paper and be sure to talk to your instructor. You might ask in class or privately with the instructor.

Do You Really Know It?

Can you put into words the concepts that you learned in this section? Answer the below question from the *Writing* section in the *Study Set* in your text. Explain as if you were explaining to someone who has never taken this class before. Use notebook paper if you need more room.

Explain how to determine from their slopes whether two lines are parallel, perpendicular, or neither.

You Write the Test!

If you were writing the test for this section, what would you want a student to know? Write two test questions that you think might come from this material. Write questions of various difficulty; these questions can be original or chosen from the homework. Be sure to supply the answer also!

Write these questions at the end of this chapter under the section titled *My Practice Chapter 2 Test* and the answer to each question under the section titled *My Practice Chapter 2 Test Answers*.

Reflect on the Section

Look back at the *Pre-Class Prep* section. Did the lecture explain topics that you thought were going to be challenging or confusing? _____

- Are there topics that you still have questions on from the reading or the lecture? If so, complete the following:
    I don't understand…

- Speak to your instructor in class or during office hours about these concerns.

Reflect on Your Math Attitude

Currently, how do you feel the class is going for you?

Considering your learning style, what might the instructor do differently to help you, or what is something that you could do differently?

2.4 Writing Equations of Lines: Pre-Class Prep

Are You Ready?

Complete the following problems. These problems review some basic skills that are needed when working with equations of lines. *All answers are found in the Pre-Class Prep Answer Section.*

1. a. Identify each term in the expression $5x - 8$.

   b. What is the coefficient of the first term?

2. Solve for y: $4x + 5y = 20$

3. On what axis does the point $(0, 3)$ lie?

4. Find the slope of the line that passes through $(-1, 0)$ and $(-10, -6)$.

5. Solve $y + 1 = 2(x - 11)$ for y.

6. Add: $-\dfrac{3}{8} + 6$

Reading Time!

While **reading Section 2.4**, identify the statements as True or False. *All answers are found in the Pre-Class Prep Answer Section.*

1 Use the Slope-Intercept Form to Write the Equation of a Line

_____ 1. The slope-intercept form of the equation of a line is $y = mx + b$.

_____ 3. The m in $y = mx + b$ represents the slope.

_____ 4. To write a linear equation in 2 variables in slope-intercept form, solve the equation for x.

_____ 5. If only the slope and y-intercept are known, an equation of the line can be written.

2 Write a Linear Equation Model

_____ 6. Equations can be written in slope-intercept form to model linear relationships.

_____ 7. In the linear model $L = -0.05t + 7.25$, the 7.25 represents the rate of change.

3 Use the Point-Slope Form to Write the Equation of a Line

_____ 8. The point-slope form of the equation of a line is $y - y_1 = m(x - x_2)$.

_____ 9. When using the point-slope form, never substitute values for x or y.

_____10. The point-slope form can be used to write the equation of a line when the slope and one point on the line are known.

4 Write a Linear Depreciation Equation

_____ 11. Straight-line depreciation and straight-line appreciation equations describe certain types of financial gain or loss.

_____ 12. An equation with slope -170 is called a line of appreciation.

5 Use Slope as an Aid When Graphing

_____ 13. If only the slope and y-intercept are known, the line can be graphed without constructing a table of values.

_____ 14. An equation in $Ax + By = C$, should be written in point-slope form before graphing.

6 Recognize Parallel and Perpendicular Lines

_____ 15. If slopes are equal, the lines are perpendicular. If the slopes are negative reciprocals, the lines are parallel.

_____ 16. A horizontal line has equation $y = b$ and slope 0.

_____ 17. The slope-intercept form is $y = mx + b$, where m is the slope and b is the y-intercept.

_____ 18. The equation of a straight line is $Ax + By = C$, where A and B cannot both be 0.

_____ 19. A vertical line has equation $x = a$, no slope, and has x-intercept $(a, 0)$.

_____ 20. The point-slope form is $y - y_1 = m(x - x_1)$, where m is the slope and the line passes through (x_1, y_1).

Getting Ready for Class

Briefly look through the section again. Answer the following by writing the concept or just the page number from the text.

Identify concepts/procedures that you feel confident about:

Identify concepts/procedures that look confusing or challenging:

Be sure to ask your instructor further questions if you are still having difficulty with a concept.

2.4 Writing Equations of Lines: In-Class Notes

Terms, Definitions, and Main Ideas	Examples and Notes

Use notebook paper for additional notes

2.4 Writing Equations of Lines: After Class

Important to Know

What is your homework assignment? Be sure to note it in your weekly schedule.

Section 2.4 Homework: _____ Due: _____

Getting to Work!

Complete your homework assignment. If you are unable to do a problem, write down the problem number and a question to help you remember what you would like to ask your instructor, your tutor, or another student.

Problem Number	Question? Where in the problem did you start to have difficulty or confusion?	Answered?

Often, you will have more questions than there is space provided here. If so, write them on notebook paper and be sure to talk to your instructor. You might ask in class or privately with the instructor.

Do You Really Know It?

Can you put into words the concepts that you learned in this section? Answer the below question from the *Writing* section in the *Study Set* in your text. Explain as if you were explaining to someone who has never taken this class before. Use notebook paper if you need more room.

Explain how to find the equation of a line passing through two given points.

You Write the Test!

If you were writing the test for this section, what would you want a student to know? Write two test questions that you think might come from this material. Write questions of various difficulty; these questions can be original or chosen from the homework. Be sure to supply the answer also!

Write these questions at the end of this chapter under the section titled *My Practice Chapter 2 Test* and the answer to each question under the section titled *My Practice Chapter 2 Test Answers*.

Reflect on the Section

Look back at the *Pre-Class Prep* section. Did the lecture explain topics that you thought were going to be challenging or confusing? _____

- Are there topics that you still have questions on from the reading or the lecture? If so, complete the following:
  I don't understand…

- Speak to your instructor in class or during office hours about these concerns.

Reflect on Your Math Attitude

Describe your feelings about your note-taking skills.

Be sure to take the time to look your notes over immediately after class, if possible. Often the instructor or another student can help fill in any missing information before you even leave the classroom!

2.5 An Introduction to Functions: Pre-Class Prep

✓ Are You Ready?

Complete the following problems. These problems review some basic skills that are needed when working with functions. *All answers are found in the Pre-Class Prep Answer Section.*

1. Which of the ordered pairs in the following set have the same *x*-coordinate:

$$(3,5),(2,9),(6,-7),(-1,5),(3,0),(1,9)$$

2. Substitute 8 for *x* in $y = \frac{1}{2}x + 3$ and find *y*.

3. What is the slope and the *y*-intercept of the line described by the equation $y = 3x - 8$?

4. What is the slope of a line perpendicular to the graph of the line that is described by the equation $y = \frac{2}{3}x + 1$?

✓ Reading Time!
(on the next page)

✓ Getting Ready for Class

Briefly look through the section again. Answer the following by writing the concept or just the page number from the text.

Identify concepts/procedures that you feel confident about:

Identify concepts/procedures that look confusing or challenging:

Be sure to ask your instructor further questions if you are still having difficulty with a concept.

Reading Time!

While **reading Section 2.5**, match the word or concept to its definition or description. *All answers are found in the Pre-Class Prep Answer Section.*

1 Define Relation, Domain, and Range

_____ 1. Relation

_____ 2. Domain of the relation

_____ 3. Range of the relation

2 Identify Functions

_____ 4. Arrow diagram and two-column tables

_____ 5. Function

_____ 6. Independent variable

3 Use Function Notation

_____ 7. Function notation

_____ 8. Function value

4 Find the Domain of a Function

_____ 9. Domain of $f(x) = 3x + 1$

_____ 10. Domain of $f(x) = \dfrac{x}{3x-6}$

5 Graph Linear Functions

_____ 11. Graph of the function

_____ 12. Graphing methods

_____ 13. Linear function

_____ 14. Identity function

6 Write Equations as Linear Functions

_____ 15. Forms to write equations of linear functions

A. x, since y depends on x

B. the input-output pairs generated by a function plotted as ordered pairs on a rectangular coordinate system

C. set of all second components in a relation

D. used to define a relation

E. most basic linear function $f(x) = x$

F. the set of all real numbers, $\mathbb{R}$, since able to evaluate $f(x)$ for any value of x

G. $y = f(x)$, the variable y is a function of x

H. point-plotting method, slope-intercept method, and intercept method

I. set of ordered pairs

J. the output of a function

K. slope-intercept form and point-slope form

L. function in the form $f(x) = mx + b$, where m and b are real numbers

M. set of ordered pairs in which to each first component there corresponds exactly one second component

N. set of all first components in a relation

O. the set of all real numbers except 2, since division by 0 is undefined

Terms, Definitions, and Main Ideas	Examples and Notes

Use notebook paper for additional notes

2.5 An Introduction to Functions: After Class

Important to Know

What is your homework assignment? Be sure to note it in your weekly schedule.

Section 2.5 Homework: _____ Due: _____

Getting to Work!

Complete your homework assignment. If you are unable to do a problem, write down the problem number and a question to help you remember what you would like to ask your instructor, your tutor, or another student.

Problem Number	Question? Where in the problem did you start to have difficulty or confusion?	Answered?

Often, you will have more questions than there is space provided here. If so, write them on notebook paper and be sure to talk to your instructor. You might ask in class or privately with the instructor.

Do You Really Know It?

Can you put into words the concepts that you learned in this section? Answer the below question from the *Writing* section in the *Study Set* in your text. Explain as if you were explaining to someone who has never taken this class before. Use notebook paper if you need more room.

Consider the function defined by $y = 6x + 4$. Why do you think x is called the independent variable and y the dependent variable?

You Write the Test!

If you were writing the test for this section, what would you want a student to know? Write two test questions that you think might come from this material. Write questions of various difficulty; these questions can be original or chosen from the homework. Be sure to supply the answer also!

Write these questions at the end of this chapter under the section titled *My Practice Chapter 2 Test* and the answer to each question under the section titled *My Practice Chapter 2 Test Answers*.

Reflect on the Section

Look back at the *Pre-Class Prep* section. Did the lecture explain topics that you thought were going to be challenging or confusing? _____

- Are there topics that you still have questions on from the reading or the lecture? If so, complete the following:
    I don't understand…

- Speak to your instructor in class or during office hours about these concerns.

Reflect on Your Math Attitude

How do you feel about your homework? That is, do you feel that you are in control or that the homework is in control of you?

Describe how knowing about your individual learning style, student-support features, and note-taking skills can help improve your attitude toward homework (even if your attitude is already good)?

2.6 Graphs of Functions: Pre-Class Prep

 Are You Ready?

Complete the following problems. These problems review some basic skills that are needed when graphing functions. *All answers are found in the Pre-Class Prep Answer Section.*

1. Fill in the blanks, $f(3) = 9$ corresponds to the ordered pair $(\_\_, \_\_)$.

2. If $f(x) = x^2$, find $f(-1)$ and $f(1)$.

3. If $f(x) = x^3$, find $f(-2)$ and $f(2)$.

4. If $f(x) = |x|$, find $f(-4)$ and $f(4)$.

 Reading Time!

While **reading Section 2.6**, identify the word or concept being defined. Choose from the following words (some may be used more than once or not at all). *All answers are found in the Pre-Class Prep Answer Section.*

cubing function	k units upward	h units to the right
h units to the left	point-plotting method	nonlinear functions
domain	absolute value function	k units downward
vertical translation	value of $f(a)$	squaring function
parabola	reflection	x-axis
y-axis	range	horizontal translation
Vertical Line Test		

Chapter 2 Graphs, Equations of Lines, and Functions

1 Find Function Values Graphically

1. Given by the y-coordinate of a point on the graph of function f with x-coordinate a: _____

2 Find the Domain and Range of a Function Graphically

2. The projection of the graph on the x-axis provides this: _____

3. The projection of the graph on the y-axis provides this: _____

3 Graph Nonlinear Functions

4. Functions whose graphs are not lines: _____

5. Method used to graph a function: _____

6. The function defined by $f(x) = x^2$: _____

7. The graph of $f(x) = x^2$: _____

8. The nonlinear function defined by $f(x) = x^3$: _____

9. The nonlinear function defined by $f(x) = |x|$: _____

4 Translate Graphs of Functions

10. Shift of a graph upward or downward: _____

11. Translation of the graph of $y = f(x) + k$ with respect to $y = f(x)$: _____

12. Translation of the graph of $y = f(x) - k$ with respect to $y = f(x)$: _____

13. Shifts of a graph to the right or to the left: _____

14. Translation of the graph of $y = f(x + h)$ with respect to $y = f(x)$: _____

15. Translation of the graph of $y = f(x - h)$ with respect to $y = f(x)$: _____

5 Reflect Graphs of Functions

16. The graph of $g(x) = -x^2$ with respect to $g(x) = x^2$: _____

17. The graph of $y = -f(x)$ is the graph of $y = f(x)$ reflected about this axis: _____

5 Use the Vertical Line Test

18. If a vertical line intersects a graph in more than one point, the graph is not the graph of a function: _____

Getting Ready for Class

Briefly look through the section again. Answer the following by writing the concept or just the page number from the text.

Identify concepts/procedures that you feel confident about:

Identify concepts/procedures that look confusing or challenging:

Be sure to ask your instructor further questions if you are still having difficulty with a concept.

Chapter 2 Graphs, Equations of Lines, and Functions

Terms, Definitions, and Main Ideas

Examples and Notes

Use notebook paper for additional notes

2.6 Graphs of Functions: After Class

Important to Know

What is your homework assignment? Be sure to note it in your weekly schedule.

Section 2.6 Homework: _____ **Due**: _____

Getting to Work!

Complete your homework assignment. If you are unable to do a problem, write down the problem number and a question to help you remember what you would like to ask your instructor, your tutor, or another student.

Problem Number	Question? Where in the problem did you start to have difficulty or confusion?	Answered?

Often, you will have more questions than there is space provided here. If so, write them on notebook paper and be sure to talk to your instructor. You might ask in class or privately with the instructor.

Do You Really Know It?

Can you put into words the concepts that you learned in this section? Answer the below question from the *Writing* section in the *Study Set* in your text. Explain as if you were explaining to someone who has never taken this class before. Use notebook paper if you need more room.

Explain how to project the graph of a function onto the x-axis. Give an example.

© 2013 Cengage Learning. All Rights Reserved. May not be scanned, copied or duplicated, or posted to a publicly accessible website, in whole or in part.

You Write the Test!

If you were writing the test for this section, what would you want a student to know? Write two test questions that you think might come from this material. Write questions of various difficulty; these questions can be original or chosen from the homework. Be sure to supply the answer also!

Write these questions at the end of this chapter under the section titled *My Practice Chapter 2 Test* and the answer to each question under the section titled *My Practice Chapter 2 Test Answers*.

Reflect on the Section

Look back at the *Pre-Class Prep* section. Did the lecture explain topics that you thought were going to be challenging or confusing? _____

- Are there topics that you still have questions on from the reading or the lecture? If so, complete the following:
    I don't understand…

- Speak to your instructor in class or during office hours about these concerns.

Reflect on Your Math Attitude

Write one word that describes how you feel about this section: _____

What could you do to either improve your current feeling to a positive attitude or to make sure that you continue to stay positive? Think about your knowledge of your learning style, student-support features, and note taking.

Chapter 2 Activities

Your instructor may assign these activities to you to complete in class, or you may complete them on your own to solidify your understanding of chapter topics. The activities begin on the next page.

❖ **Student Activity:** *The Cost of College*
   Create a linear model and graph of college costs.

❖ **Student Activity:** *Entertaining Rates of Change*
   Summarize raw data with a rate describing the amount of change in one quantity with respect to the amount of change in another.

Student Activity
The Cost of College

When you pay for your college education, there are typically three different types of charges:

1) *Variable cost* – Tuition, which is paid per credit hour (or unit of study). There may also be fees that vary depending on the number of credit hours (for example, an additional $6 per credit hour).
2) *Fixed cost* – Student fees, which are paid regardless of how many or what types of classes you take, (for example, a $25 Registration Fee). These are fixed costs.
3) Course fees or lab fees, which are paid for certain courses only (we will ignore these in this model). Your instructor may tell you to ignore some other fee or tuition oddities to allow a linear model to be used.

Our linear model of college costs will only consider the first two types of charges. Your instructor will help you locate the information necessary to build the linear model – perhaps they have asked you to bring in a copy of your tuition bill!

How much do you pay in **fixed** costs for your enrollment at your college? _____

Per credit hour, how much do you pay in tuition (and **variable** fees)? _____

Fill in the table below for a student at your college:

Credits (or units) taken by student, n	Fixed student fees	Tuition and other variable costs for this number of credits (or units)	Total cost for this number of credits, C
4			
8			
12			
16			

Create a graph containing the data points above.

Write a linear equation to model the cost C, for taking a certain number of credits n.

Show how you could use the linear equation you just wrote to estimate the cost of taking 9 credits (or units).

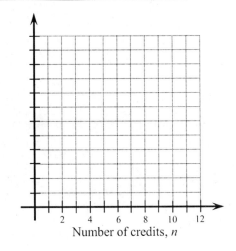

Total Cost

Number of credits, *n*

Student Activity
Entertaining Rates of Change

It is often helpful to summarize raw data with a rate describing the amount of change in one quantity with respect to the amount of change in another. This is called an **average rate of change**, and is often used to describe change that occurs *over time*.

$$\text{Average Rate of Change} = \frac{\text{change in quantity}}{\text{change in time}}$$

Directions: Answer the questions below by calculating an average rate of change and using this information to answer the follow-up questions.

1. This table contains box office data from 2000 – 2006. (Source: www.boxofficemojo.com)

Year	Total Gross (in millions)	Tickets Sold (in millions)	# of Pics	Ticket Price	#1 Picture of the year
2006	$9,209.4	1400.0	606	$6.58	Dead Man's Chest
2005	$8,840.4	1381.3	547	$6.40	Revenge of the Sith
2004	$9,418.3	1516.6	551	$6.21	Shrek 2
2003	$9,185.9	1523.3	508	$6.03	Return of the King
2002	$9,167.0	1578.0	467	$5.81	Spider-Man
2001	$8,412.5	1487.3	482	$5.66	Harry Potter
2000	$7,661.0	1420.8	478	$5.30	The Grinch

a. What is the average annual rate of change for the total box office gross between 2000 and 2006? Show your calculation and round your answer to the nearest tenth.

b. Complete the statement: *On average, total box office gross increased _____ dollars per year over the 6-year period between 2000 and 2006. The total increase in box office gross during this period was _____.*

c. What is the average annual rate of change for the total box office gross between 2003 and 2006? Show your calculation and round your answer to the nearest tenth.

d. Complete the statement: *On average, total box office gross increased _____ dollars per year over the 3-year period between 2003 and 2006. The total increase in box office gross during this period was _____.*

e. In general, Is the average annual rate of change for total box office gross increasing or decreasing? _____

The following tables contain U.S. Music purchasing data for 2005 and 2006.

Units sold (in millions):	2005	2006
Overall Music Sales (Albums, singles, music video, digital tracks)	1,003	1,198
Total Album Sales (Includes CD, CS, LP, Digital albums)	618.9	588.2
Digital Track Sales	352.7	581.9
Overall Album Sales (Includes all albums & track equivalent albums)	654.1	646.4
Internet Album Sales (Physical album purchases via e-commerce sites)	24.7	29.4
Digital Album Sales	16.2	32.6

(Source: www.businesswire.com)

Top Ten Selling Albums of 2006	Units Sold
Soundtrack/High School Musical	3,719,071
Me and My Gang/Rascal Flatts	3,479,994
Some Hearts/Carrie Underwood	3,015,950
All the Right Reasons/Nickelback	2,688,166
Futuresex/Lov…/Justin Timberlake	2,377,127
Back to Bedlam/James Blunt	2,137,142
B'day/Beyonce	2,010,311
Soundtrack/Hannah Montana	1,987,681
Taking the Long Way/Dixie Chicks	1,856,284
Extreme Behavior/Hinder	1,817,350

2. Use the tables above to answer the following questions. Show your calculations and round your answer to the nearest hundredth.

    a. What is the average *monthly* rate at which **overall music sales** increased from 2005 to 2006?

    b. What is the average *monthly* rate at which **total album sales** decreased from 2005 to 2006?

    c. What is the average *monthly* rate at which **digital album sales** increased from 2005 to 2006?

3. Based on your results from parts b and c, does it appear that all the people who topped buying physical albums are now purchasing digital albums? Explain.

4. What is the average *daily* sales rate for the top selling album of 2006? What is the average daily sales rate for the *tenth* top selling album of 2006? (Round your answer to the nearest whole number.)

Chapter 2 Test Skills Assessment

Pre-Test Preparation Work:

1. Re-read the objectives from each section.

2. Review the *Reading Time!* activity for each section.

3. Go over all your classroom notes, if something in your notes doesn't make sense to you, make a note and ask your teacher or a classmate.

4. Make additional notations to your work if your teacher states specific concepts to study in preparation for the chapter test.

5. Attend any study sessions held by your teacher or teaching assistant.

6. Practice additional problems.

7. Go over any missed problems in your homework sets.

8. Talk out concepts with your peers in small group study sessions.

List other preparations that you have found beneficial in preparing for a math test.

Additional Practice Suggestions

1. Use your book's review problems at the back of the chapter as a practice test.

2. Take *My Practice Chapter Test* and the text's *Chapter Test*. Time yourself and do not use your notebook or textbook.

3. Pace yourself as you work through these problems.

4. Read each question carefully, playing close attention to the instructions.

5. Check your work using the answers provided.

6. Rework any missed problems. Do not just "look them over" but actually rework the problem without looking at text or notes.

My Practice Chapter 2 Test

For each section, you had the opportunity to create two test questions under the section *You Write the Test!* Write each of those questions here. Include your answers under the heading *My Practice Chapter 2 Test Answers*. **Take the test without notes or your textbook.** If you do not get a question correct, review the text and/or your notes then take the test again. For further review, do the *Chapter 2 Test* in the text.

Section 2.1

1.

2.

Section 2.2

3.

4.

Section 2.3

5.

6.

Section 2.4

7.

8.

Section 2.5

9.

10.

Section 2.6

11.

12.

My Practice Chapter 2 Test Answers

Section 2.1

1.

2.

Section 2.2

3.

4.

Section 2.3

5.

6.

Section 2.4

7.

8.

Section 2.5

9.

10.

Section 2.6

11.

12.

Chapter 3 Systems of Equations

Read the *Study Skills Workshop* found at the beginning of Chapter 3 in your textbook. **Complete** the activities below for this chapter's *Study Skills Workshop*.

Successful Test Taking

Below are some suggestions that can make taking a test more enjoyable and also improve your score.

Preparing For the Text

Studying several days before the test rather than cramming your studying into one marathon session the night before is a more effective way of preparing for an exam. A study session plan is a helpful tool.

✓ Write your study session plan below. For some suggestions, see *Preparing for a Test**.

Days Before Test	What I plan to do in order to prepare for the test.
4	
3	
2	
1	
Test Day	

Taking the Test

There are strategies to taking tests effectively. These test-taking strategies can maximize your score by using the testing time wisely.

✓ Complete the survey questions found in *How to Take a Math Test**.
✓ Summarize your test-taking strategy.

Evaluating Your Performance

When you receive your graded test, take time to study the types of errors that you made on the test so that you do not do them again.

✓ Use the outline found in *Analyzing Your Test Results** to classify the errors that you made on your most recent test.
✓ What was your most common error?

*Found online at: www.cengage.com/math/tussy

Chapter 3 Systems of Equations

© 2013 Cengage Learning. All Rights Reserved. May not be scanned, copied or duplicated, or posted to a publicly accessible website, in whole or in part.

Are You Ready?

Complete the following problems. These review some basic skills that are needed when solving systems of equations by graphing. *All answers are found in the Pre-Class Prep Answer Section.*

1. Is $(-1,4)$ a solution of $y = 3x - 1$?

2. Use the slope and the y-intercept to graph $y = -4x + 2$.

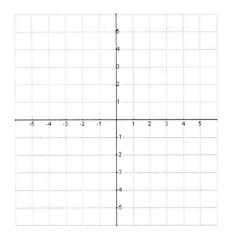

3. Graph $3x + 4y = 12$ by finding the x- and y-intercepts.

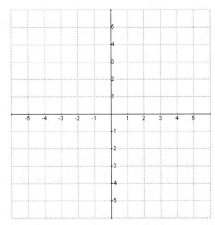

4. Without graphing, determine whether the graphs of $y = \frac{1}{3}x - 2$ and $x - 3y = 2$ are

   parallel, perpendicular, or neither.

Reading Time!

While **reading Section 3.1**, fill in the blanks choosing from the following words (some may be used more than once or not at all). *All answers are found in the Pre-Class Prep Answer Section.*

solution	inconsistent	y	solution of the system	one
system of equations	graphing	graph	equivalent systems	same
linear equations	solve	intersect	dependent	check
no	consistent	independent	x	different

1 Determine Whether a Given Ordered Pair is a Solution of a System

1. When two equations with the same variables are considered simultaneously, they form a

    _____ .

2. An ordered pair that satisfies both equations in a system of equations is called a _____ .

2 Solve Systems of Linear Equations by Graphing

3. To _____ a system of equations means to find all of the solutions of the system.

4. To use the _____ method, graph both equations on the _____ rectangular coordinate system.

5. If the lines _____ , the coordinates of the point of intersection is the _____ of the system.

6. If the graphs have no point in common, the system has _____ solution.

7. _____ the proposed solution in each equation of the system.

8. A(n) _____ system is a system of equations that has at least one solution.

3 Use Graphing to Identify Inconsistent Systems and Dependent Equations

9. Equations of a system with different graphs are _____ equations.

10. A(n) _____ system is a system of equations with no solution.

11. Equations with the same graphs are _____ equations.

12. _____ are systems where each equation in one system is equivalent to a corresponding equation in another system.

Chapter 3 Systems of Equations

4 Solve Equations Graphically

13. The graphing method can be used to solve _____ in one variable.

14. To find the solution of $2x + 4 = -2$, set $y = 2x + 4$ and $y = -2$, then solve the system by _____. The solution is the value of _____ that makes $2x + 4$ equal -2.

15. A _____ calculator can be used to solve systems of equations.

Getting Ready for Class

Briefly look through the section again. Answer the following by writing the concept or just the page number from the text.

Identify concepts/procedures that you feel confident about:

Identify concepts/procedures that look confusing or challenging:

Be sure to ask your instructor further questions if you are still having difficulty with a concept.

3.1 Solving Systems of Equations by Graphing: In-Class Notes

Terms, Definitions, and Main Ideas	**Examples and Notes**

Use notebook paper for additional notes

3.1 Solving Systems of Equations by Graphing: After Class

Important to Know

What is your homework assignment? Be sure to note it in your weekly schedule.

Section 3.1 Homework: _____ **Due**: _____

Getting to Work!

Complete your homework assignment. If you are unable to do a problem, write down the problem number and a question to help you remember what you would like to ask your instructor, your tutor, or another student.

Problem Number	Question? Where in the problem did you start to have difficulty or confusion?	Answered?

Often, you will have more questions than there is space provided here. If so, write them on notebook paper and be sure to talk to your instructor. You might ask in class or privately with the instructor.

Do You Really Know It?

Can you put into words the concepts that you learned in this section? Answer the below question from the *Writing* section in the *Study Set* in your text. Explain as if you were explaining to someone who has never taken this class before. Use notebook paper if you need more room.

Can a system of two linear equations have exactly two solutions? Why or why not?

You Write the Test!

If you were writing the test for this section, what would you want a student to know? Write test questions that you think might come from this material. Write questions of various difficulty; these questions can be original or chosen from the homework. Be sure to supply the answer also!

Write these questions at the end of this chapter under the section titled *My Practice Chapter 3 Test* and the answer to each question under the section titled *My Practice Chapter 3 Test Answers*.

Reflect on the Section

Look back at the *Pre-Class Prep* section. Did the lecture explain topics that you thought were going to be challenging or confusing? _____

- Are there topics that you still have questions on from the reading or the lecture? If so, complete the following:
    I don't understand…

- Speak to your instructor in class or during office hours about these concerns.

Reflect on Your Math Attitude

What do you think about graphing? Do you feel confident in your skills? _____

If you do not feel confident, you might want to seek out help. This could be your instructor, your tutor, or another student.

Whom would **you** seek help from? _____

If you do feel confident, helping another student is a great way to make sure that you do know the material. Are there any other students that could use your help?

If so, who?_____

Are You Ready?

Complete the following problems. These review some basic skills that are needed when solving systems of equations algebraically. *All answers are found in the Pre-Class Prep Answer Section.*

1. In $6x + y = 9$, what is the coefficient of y?

2. Solve $5x - y = -4$ for y.

3. Substitute 4 for x in $y = -2x - 1$ and find y.

4. In $8x - 3y = 12$, what is the *opposite* of the coefficient of y?

5. Substitute -3 for y in $5x + 6y = 2$ and find x.

6. Multiply both sides of the equation $7x - y = 9$ by -4.

Reading Time!

While **reading Section 3.2**, identify the word or concept being defined. Choose from the following words (some may be used more than once or not at all). *All answers are found in the Pre-Class Prep Answer Section.*

consistent	no solution	infinitely many solutions
terms involving x (or y)	$Ax + By = C$	substitution method
elimination method	inconsistent	step 1 of the substitution method
substitution equation	opposites	step 2 of the substitution method
step 4 of the substitution method	$A + B = C$	step 3 of the substitution method
substitution method	graphing method	elimination method

1 Solve Systems of Linear Equations by Substitution

1. One algebraic method for solving a system of equations: _____

2. The equation used to make a substitution: _____

3. Substitute the value of the variable found into the substitution equation to find the value of the remaining variable: _____

4. Substitute the expression for x or for y obtained earlier into the other equation and solve that equation: _____

5. Solve one of the equations for either x or y: _____

6. Check the proposed solution in each equation of the original system: _____

2 Solve Systems of Linear Equations by the Elimination (Addition) Method

7. A method based on using the addition property of equality to solve a system: _____

8. Equations in the system must be written in this standard form: _____

9. Multiply one or both of the equations by a nonzero number chosen to make the coefficients of x (or the coefficients of y) this: _____

10. Equations are added to eliminate one of these: _____

3 Use Substitution and Elimination (Addition) to Identify Inconsistent Systems and Dependent Equations

11. The solution of a system that resulted in an identity (true statement): _____

12. The solution of a system that resulted in a contradiction (false statement): _____

13. A system of equations with no solutions: _____

14. A system of equations with an infinite number of solutions: _____

4 Determine the Most Efficient Method to Use to Solve a Linear System

15. Most efficient method when one of the equations is solved for one of the variables: _____

16. Most efficient method when both equations are in standard $Ax + By = C$ form and no variable has a coefficient of 1 or −1: _____

17. Most efficient method when trends need to be shown or to see the point that two graphs have in common: _____

Getting Ready for Class

Briefly look through the section again. Answer the following by writing the concept or just the page number from the text.

Identify concepts/procedures that you feel confident about:

Identify concepts/procedures that look confusing or challenging:

Be sure to ask your instructor further questions if you are still having difficulty with a concept.

3.2 Solving Systems of Equations Algebraically: In-Class Notes

Terms, Definitions, and Main Ideas	Examples and Notes

Use notebook paper for additional notes

3.2 Solving Systems of Equations Algebraically: After Class

 Important to Know

What is your homework assignment? Be sure to note it in your weekly schedule.

Section 3.2 Homework: _____ **Due**: _____

Getting to Work!

Complete your homework assignment. If you are unable to do a problem, write down the problem number and a question to help you remember what you would like to ask your instructor, your tutor, or another student.

Problem Number	Question? Where in the problem did you start to have difficulty or confusion?	Answered?

Often, you will have more questions than there is space provided here. If so, write them on notebook paper and be sure to talk to your instructor. You might ask in class or privately with the instructor.

Do You Really Know It?

Can you put into words the concepts that you learned in this section? Answer the below question from the *Writing* section in the *Study Set* in your text. Explain as if you were explaining to someone who has never taken this class before. Use notebook paper if you need more room.

Why is the method for solving systems that is discussed in this section called the elimination method*?*

You Write the Test!

If you were writing the test for this section, what would you want a student to know? Write two test questions that you think might come from this material. Write questions of various difficulty and these questions can be original or chosen from the homework. Be sure to supply the answer also!

Write these questions at the end of this chapter under the section titled *My Practice Chapter 3 Test* **and the answer to each question under the section titled** *My Practice Chapter 3 Test Answers.*

Reflect on the Section

Look back at the *Pre-Class Prep* section. Did the lecture explain topics that you thought were going to be challenging or confusing? _____

- Are there topics that you still have questions on from the reading or the lecture? If so, complete the following:
    I don't understand…

- Speak to your instructor in class or during office hours about these concerns.

Reflect on Your Math Attitude

Did you have an exam over the last chapter? If so, how did you feel about the test when you completed it?

Be sure to classify errors. What are your thoughts about the type of errors that you made?

3.3 Solving Systems of Equations in Three Variables: Pre-Class Prep

Are You Ready?

Complete the following problems. These problems review some basic skills that are needed when solving systems of equations in three variables. *All answers are found in the Pre-Class Prep Answer Section.*

1. What is the coefficient of each term on the left side of the equation $x - 4y + 5z = 2$?

2. For $6x + y + 2z = 36$, if $x = 5$ and $y = -2$, what is z?

3. Write the equation $7x + 5y = z + 9$ so that all three variable terms are on the left side.

4. Solve the system by elimination: $\begin{cases} 5x + 2y = -5 \\ -6x - 2y = 10 \end{cases}$

Reading Time!

While **reading Section 3.3**, identify the statements as True or False. *All answers are found in the Pre-Class Prep Answer Section.*

1 Determine Whether an Ordered Triple is a Solution of a System

_____ 1. A linear equation in three variables is an equation that can be written in the form $Ax + By + Cz = D$, where $A, B, C,$ and D are real numbers and $A, B,$ and C are not all 0.

_____ 2. A solution of a linear equation in three variables in an ordered pair of numbers of the form (x, y, z), whose coordinates satisfy the equation.

_____ 3. A solution of a system of three linear equations in three variables is an ordered triple that satisfies each equation of the system.

_____ 4. The graph of an equation of the form $Ax + By + Cz = D$ is a plane.

_____ 5. Three planes with a line l in common is a system with one solution.

2 Solve Systems of Three Linear Equations in Three Variables

Steps to Solving a System of Three Equations

_____ 6. Write each equation in standard form $Ax + By = C$ and clear any decimals or fractions.

_____ 7. Pick any two equations and eliminate a variable.

_____ 8. Pick a different pair of equations and eliminate the same variable as before.

_____ 9. Solve the resulting pair of three equations in three variables.

_____ 10. Substitute values of the two variables found into any equation containing all three variables and solve the equation to find the value of the third variable.

_____ 11. Check the solution in one of the original equations, then write the solution as an ordered triple.

3 Solve Systems of Equations with Missing Variable Terms

_____ 12. If a variable term is missing, the elimination of a variable that is normally performed in step 2 of the solution process is skipped.

_____ 13. When finding the values of the variables, find them in alphabetical order.

4 Identify Inconsistent Systems and Dependent Equations

_____ 14. If the planes of a system of three equations have common point of intersection, the system is said to be consistent.

_____ 15. A consistent and dependent system can be a system of equations such that the three planes coincide or that the three planes intersect to form a common line.

Getting Ready for Class

Briefly look through the section again. Answer the following by writing the concept or just the page number from the text.

Identify concepts/procedures that you feel confident about:

Identify concepts/procedures that look confusing or challenging:

Be sure to ask your instructor further questions if you are still having difficulty with a concept.

3.3 Solving Systems of Equations in Three Variables: In-Class Notes

Terms, Definitions, and Main Ideas	Examples and Notes

Use notebook paper for additional notes

3.3 Solving Systems of Equations in Three Variables: After Class

Important to Know

What is your homework assignment? Be sure to note it in your weekly schedule.

Section 3.3 Homework: _____ **Due**: _____

Getting to Work!

Complete your homework assignment. If you are unable to do a problem, write down the problem number and a question to help you remember what you would like to ask your instructor, your tutor, or another student.

Problem Number	Question? Where in the problem did you start to have difficulty or confusion?	Answered?

Often, you will have more questions than there is space provided here. If so, write them on notebook paper and be sure to talk to your instructor. You might ask in class or privately with the instructor.

Do You Really Know It?

Can you put into words the concepts that you learned in this section? Answer the below question from the *Writing* section in the *Study Set* in your text. Explain as if you were explaining to someone who has never taken this class before. Use notebook paper if you need more room.

What makes a system of three equations with three variables inconsistent?

You Write the Test!

If you were writing the test for this section, what would you want a student to know? Write two test questions that you think might come from this material. Write questions of various difficulty; these questions can be original or chosen from the homework. Be sure to supply the answer also!

Write these questions at the end of this chapter under the section titled *My Practice Chapter 3 Test* and the answer to each question under the section titled *My Practice Chapter 3 Test Answers*.

Reflect on the Section

Look back at the *Pre-Class Prep* section. Did the lecture explain topics that you thought were going to be challenging or confusing? _____

- Are there topics that you still have questions on from the reading or the lecture? If so, complete the following:
    I don't understand…

- Speak to your instructor in class or during office hours about these concerns.

Reflect on Your Math Attitude

Did this topic of solving systems of three equations make sense to you? If so or if not, how does that make you feel?

What can you do to maintain a positive attitude, or develop a positive attitude, about this new concept?

When you begin your homework, be sure to review your notes and the examples in your text.

Chapter 3 Systems of Equations

3.4 Solving Systems of Equations Using Matrices: Pre-Class Prep

 Are You Ready?

Complete the following problems. These review some basic skills that are needed when solving systems of equations using matrices. *All answers are found in the Pre-Class Prep Answer Section.*

1. Consider the system $\begin{cases} 3x - 7y = 14 \\ 5x - y = -9 \end{cases}$. What are the coefficients of the variable terms on the left side of

   a. the first equation?

   b. the second equation?

2. Multiply -2 times 4 and add the result to 9. What is the answer?

3. Multiply -18 by $-\dfrac{1}{18}$. What is the result?

4. Multiply both sides of $2x - 6y = 5$ by -2. What is the result?

 Reading Time!

While **reading Section 3.4**, fill in the blanks choosing from the following words (some may be used more than once or not at all). *All answers are found in the Pre-Class Prep Answer Section.*

columns	horizontal	constants	main diagonal	interchanged
no	coefficients	element	matrix	adding
augmented matrix	rows	0's	one equation	dependent
vertical	infinitely many	nonzero	reduced row-echelon form	solution
order	original	inconsistent	Gauss-Jordan elimination	1's

1 Define a Matrix and Determine Its Order

1. A _____ is any rectangular array of numbers arranged in _____ and columns, written within brackets.

2. The rows of a matrix are _____ and the columns are _____.

3. Each number in a matrix is called an _____ or an entry of the matrix.

4. A matrix with m rows and n columns has _____ $m \times n$, which is read as "m by n."

2 Write the Augmented Matrix for a System

5. A system of two linear equations $\begin{cases} x - y = 4 \\ 2x + y = 5 \end{cases}$ would have the _____ $\begin{bmatrix} 1 & -1 & |4 \\ 2 & 1 & |5 \end{bmatrix}$.

6. Each row of the augmented matrix represents _____ of the system. The first two columns are determined by the _____ of x and y and the last column is determined by the _____ in the equations.

3 Perform Elementary Row Operations on Matrices

7. To solve a system of equations, transform the _____ into an equivalent matrix with _____ down its _____ and 0's in all the remaining entries directly above and below this diagonal.

8. A matrix written in the form described in question 7 is said to be in _____.

9. *Elementary Row Operations*:

   Type 1: Any two rows of a matrix can be _____.

   Type 2: Any row of a matrix can be multiplied by a _____ constant.

   Type 3: Any row of a matrix can be changed by _____ a nonzero constant multiple of another row to it.

4 Use Matrices to Solve a System of Two Equations

10. The process _____ uses a series of elementary row operations on an augmented matrix to produce a simpler, equivalent matrix.

11. *Solving Systems of Linear Equations Using Gauss-Jordan Elimination*

   1. Write an _____ for the system.

   2. Use elementary row operations to transform the augmented matrix into a matrix in ____.

   3. When step 2 is complete, write the resulting equivalent system to find the _____.

   4. Check the proposed solutions in the equations of the _____ system.

5 Use Matrices to Solve a System of Three Equations

12. Use _____ to solve a system of three equations.

6 Use Matrices to Identify Inconsistent Systems and Dependent Equations

13. The augmented matrix $\begin{bmatrix} 1 & 1 & | -1 \\ 0 & 0 & | -8 \end{bmatrix}$ indicates that the system is _____ and has _____ solution(s).

14. The augmented matrix $\begin{bmatrix} 1 & -\dfrac{1}{2} & | 2 \\ 0 & 0 & | 0 \end{bmatrix}$ indicates that the equations are _____ and the system

has _____ solution(s).

✓ Getting Ready for Class

Briefly look through the section again. Answer the following by writing the concept or just the page number from the text.

Identify concepts/procedures that you feel confident about:

Identify concepts/procedures that look confusing or challenging:

Be sure to ask your instructor further questions if you are still having difficulty with a concept.

3.4 Solving Systems of Equations Using Matrices: In-Class Notes

Terms, Definitions, and Main Ideas	Examples and Notes

Use notebook paper for additional notes

3.4 Solving Systems of Equations Using Matrices: After Class

Important to Know

What is your homework assignment? Be sure to note it in your weekly schedule.

Section 3.4 Homework: _____ **Due**: _____

Getting to Work!

Complete your homework assignment. If you are unable to do a problem, write down the problem number and a question to help you remember what you would like to ask your instructor, your tutor, or another student.

Problem Number	Question? Where in the problem did you start to have difficulty or confusion?	Answered?

Often, you will have more questions than there is space provided here. If so, write them on notebook paper and be sure to talk to your instructor. You might ask in class or privately with the instructor.

Do You Really Know It?

Can you put into words the concepts that you learned in this section? Answer the below question from the *Writing* section in the *Study Set* in your text. Explain as if you were explaining to someone who has never taken this class before. Use notebook paper if you need more room.

Explain how a type 3 row operation is similar to the elimination (addition) method of solving a system of equations.

You Write the Test!

If you were writing the test for this section, what would you want a student to know? Write test questions that you think might come from this material. Write questions of various difficulty; these questions can be original or chosen from the homework. Be sure to supply the answer also!

Write these questions at the end of this chapter under the section titled *My Practice Chapter 3 Test* and the answer to each question under the section titled *My Practice Chapter 3 Test Answers*.

Reflect on the Section

Look back at the *Pre-Class Prep* section. Did the lecture explain topics that you thought were going to be challenging or confusing? _____

- Are there topics that you still have questions on from the reading or the lecture? If so, complete the following:
    I don't understand…

- Speak to your instructor in class or during office hours about these concerns.

Reflect on Your Math Attitude

How do you feel about your homework? Describe what has or has not worked for you with respect to your homework. Would you make any changes?

Keeping your homework neat and organized will help when you start preparing for your test.

Are You Ready?

Complete the following problems. These review some basic skills that are needed when solving systems of equations using determinants. *All answers are found in the Pre-Class Prep Answer Section.*

1. How many rows and columns does the matrix $\begin{bmatrix} 2 & -4 & 9 \\ 1 & 0 & 4 \\ -6 & 5 & 11 \end{bmatrix}$ have?

2. What numbers lie on the main diagonal of the matrix $\begin{bmatrix} 3 & 5 \\ 6 & -1 \end{bmatrix}$?

3. Evaluate: $6(-8)-(-2)(-3)$

4. Evaluate: $6\left[3-(-1)\right]-9\left[7-(-8)\right]+4(-2-1)$

Reading Time!

While **reading Section 3.5**, identify the word or concept being defined. Choose from the following words/expressions (some may be used more than once or not at all). *All answers are found in the Pre-Class Prep Answer Section.*

$a_1 \begin{vmatrix} b_2 & c_2 \\ b_3 & c_3 \end{vmatrix} + b_1 \begin{vmatrix} a_2 & c_2 \\ a_3 & c_3 \end{vmatrix} - c_1 \begin{vmatrix} a_2 & b_2 \\ a_3 & b_3 \end{vmatrix}$

D_y

$ab - dc$

$\begin{matrix} + & - & + \\ - & + & - \\ + & - & + \end{matrix}$

determinant

Cramer's rule square matrix 0 (zero)

expanding by minors

$ad - bc$

$x = \dfrac{D_x}{D}, \ y = \dfrac{D_y}{D}$

$x = \dfrac{D_x}{D}, \ y = \dfrac{D_y}{D}, z = \dfrac{D_z}{D}$

$a_1 \begin{vmatrix} b_2 & c_2 \\ b_3 & c_3 \end{vmatrix} - b_1 \begin{vmatrix} a_2 & c_2 \\ a_3 & c_3 \end{vmatrix} + c_1 \begin{vmatrix} a_2 & b_2 \\ a_3 & b_3 \end{vmatrix}$

$\begin{matrix} - & + & - \\ + & - & + \\ - & + & - \end{matrix}$

$x = \dfrac{D}{D_x}, \ y = \dfrac{D}{D_y}$

$x = \dfrac{D}{D_x}, \ y = \dfrac{D}{D_y}, z = \dfrac{D}{D_z}$

inconsistent

consistent D

D_x

1 Evaluate 2 × 2 and 3 × 3 Determinants

1. A matrix that has the same number of rows and columns: _____

2. A number associated with a square matrix: _____

3. The value of the determinant of the matrix $\begin{bmatrix} a & b \\ c & d \end{bmatrix}$: _____

4. How a 3 × 3 determinant is evaluated: _____

5. Value of the 3 × 3 determinant $\begin{vmatrix} a_1 & b_1 & c_1 \\ a_2 & b_2 & c_2 \\ a_3 & b_3 & c_3 \end{vmatrix}$: _____

6. Array of signs for a 3 × 3 determinant: _____

2 Use Cramer's Rule to Solve Systems of Two Equations

7. Method of using determinants to solve systems of linear equations: _____

8. The solution of the system $\begin{cases} ax + by = e \\ cx + dy = f \end{cases}$: _____

9. The value of determinants D, D_x, and D_y if the system is consistent, but the equations dependent: _____

10. If $D = 0$ and D_x or D_y is nonzero, the system is this: _____

11. If $D \neq 0$, then the equations are independent and the system is this: _____

3 Use Cramer's Rule to Solve Systems of Three Equations

12. The solution of the system $\begin{cases} ax + by + cz = j \\ dx + cy + fz = k \\ gx + by + iz = l \end{cases}$: _____

13. Determinant that only uses the coefficients of the variables: _____

14. Determinant that replaces the y-term coefficients with the constants: _____

Chapter 3 Systems of Equations

Getting Ready for Class

Briefly look through the section again. Answer the following by writing the concept or just the page number from the text.

Identify concepts/procedures that you feel confident about:

Identify concepts/procedures that look confusing or challenging:

Be sure to ask your instructor further questions if you are still having difficulty with a concept.

3.5 Solving Systems of Equations Using Determinants: In-Class Notes

Terms, Definitions, and Main Ideas	Examples and Notes

Use notebook paper for additional notes

3.5 Solving Systems of Equations Using Determinants: After Class

Important to Know

What is your homework assignment? Be sure to note it in your weekly schedule.

Section 3.5 Homework: _____ **Due**: _____

Getting to Work!

Complete your homework assignment. If you are unable to do a problem, write down the problem number and a question to help you remember what you would like to ask your instructor, your tutor, or another student.

Problem Number	Question? Where in the problem did you start to have difficulty or confusion?	Answered?

Often, you will have more questions than there is space provided here. If so, write them on notebook paper and be sure to talk to your instructor. You might ask in class or privately with the instructor.

Do You Really Know It?

Can you put into words the concepts that you learned in this section? Answer the below question from the *Writing* section in the *Study Set* in your text. Explain as if you were explaining to someone who has never taken this class before. Use notebook paper if you need more room.

Explain the difference between a matrix and a determinant. Give an example of each.

You Write the Test!

If you were writing the test for this section, what would you want a student to know? Write two test questions that you think might come from this material. Write questions of various difficulty and these questions can be original or chosen from the homework. Be sure to supply the answer also!

Write these questions at the end of this chapter under the section titled *My Practice Chapter 3 Test* and the answer to each question under the section titled *My Practice Chapter 3 Test Answers*.

Reflect on the Section

Look back at the *Pre-Class Prep* section. Did the lecture explain topics that you thought were going to be challenging or confusing? _____

- Are there topics that you still have questions on from the reading or the lecture? If so, complete the following:
    I don't understand…

- Speak to your instructor in class or during office hours about these concerns.

Reflect on Your Math Attitude

Hopefully, you were able to read *Preparing for a Test**. What did you think about the suggestions?

Which suggestions will you apply as you prepare for your next test?

After you take your next test or quiz, answer this question: Are you pleased or disappointed with the results of applying the suggestions?

Be sure to review *Preparing for a Test** as you get closer to your next test.
*Found online at: www.cengage.com/math/tussy

3.6 Problem Solving Using Systems of Two and Three Equations: Pre-Class Prep

Are You Ready?

Complete the following problems. These problems review some basic skills that are needed when solving application problems using systems of two equations. *All answers are found in the Pre-Class Prep Answer Section.*

1. Translate to mathematical symbols: 150 less than the weight w.

2. At $260 per ounce, what is the value of four ounces of Chanel No. 5 perfume?

3. At 12 miles per hour, how far will a bicyclist travel in 6 hours?

4. An 80 pound bag of quick-set concrete is 40% sand by weight. How many pounds are in the bag?

Reading Time!
(on the next page)

Getting Ready for Class

Briefly look through the section again. Answer the following by writing the concept or just the page number from the text.

Identify concepts/procedures that you feel confident about:

Identify concepts/procedures that look confusing or challenging:

Be sure to ask your instructor further questions if you are still having difficulty with a concept.

Reading Time!

While **reading Section 3.6**, match the word or concept to its definition or description. *All answers are found in the Pre-Class Prep Answer Section.*

1 Assign Variables to Two Unknowns

_____ 1. Analyze the problem

_____ 2. Assign variables

_____ 3. Form a system of equations

_____ 4. Solve the system

_____ 5. State the conclusion

_____ 6. Check the results

2 Use Systems to Solve Geometry Problems

_____ 7. Parallelogram

_____ 8. Opposite angles of a parallelogram

_____ 9. Alternate interior angles

3 Use Systems to Solve Number-Value Problems

_____ 10. Total value

4 Use Systems to Find the Break Point

_____ 11. Setup costs

_____ 12. Unit costs

_____ 13. Break point

5 Use Systems to Solve Interest, Uniform Motion, and Mixture Problems

_____ 14. Interest formula

_____ 15. Speed of blimp with the wind

_____ 16. Speed of blimp against the wind

_____ 17. Amount of pure acid

_____ 18. Total value

A. angles created when a diagonal intersects two parallel sides of a parallelogram; have the same measure

B. $I = Prt$

C. amount of solution · strength of solution

D. represent the unknown values in the problem

E. $s + w$

F. use the words of the problem, not the equations formed in step 3

G. understand the given facts by careful reading

H. have the same measure

I. four-sided figure with two pairs of parallel sides

J. number · value

K. clearly and include units in the problem

L. cost to make one item

M. amount · price

N. translate words of the problem into mathematical symbols

O. use graphing, substitution, elimination, matrices, or Cramer's rule

P. number of items to be produced that will cost equal amounts using either machine

Q. includes costs of setting up a machine to do a certain job

R. $s - w$

Chapter 3 Systems of Equations

3.6 Problem Solving Using Systems of Two and Three Equations: In-Class Notes

Terms, Definitions, and Main Ideas	Examples and Notes

Use notebook paper for additional notes

3.6 Problem Solving Using Systems of Two and Three Equations: After Class

Important to Know

What is your homework assignment? Be sure to note it in your weekly schedule.

Section 3.6 Homework: _____ **Due**: _____

Getting to Work!

Complete your homework assignment. If you are unable to do a problem, write down the problem number and a question to help you remember what you would like to ask your instructor, your tutor, or another student.

Problem Number	Question? Where in the problem did you start to have difficulty or confusion?	Answered?

Often, you will have more questions than there is space provided here. If so, write them on notebook paper and be sure to talk to your instructor. You might ask in class or privately with the instructor.

Do You Really Know It?

Can you put into words the concepts that you learned in this section? Answer the below question from the *Writing* section in the *Study Set* in your text. Explain as if you were explaining to someone who has never taken this class before. Use notebook paper if you need more room.

To solve mixture problems, do you prefer the one-variable or two-variable solution strategy? Explain why.

You Write the Test!

     If you were writing the test for this section, what would you want a student to know? Write two test questions that you think might come from this material. Write questions of various difficulty; these questions can be original or chosen from the homework. Be sure to supply the answer also!

Write these questions at the end of this chapter under the section titled *My Practice Chapter 3 Test* and the answer to each question under the section titled *My Practice Chapter 3 Test Answers*.

Reflect on the Section

     Look back at the *Pre-Class Prep* section. Did the lecture explain topics that you thought were going to be challenging or confusing? _____

- Are there topics that you still have questions on from the reading or the lecture? If so, complete the following:
  I don't understand…

- Speak to your instructor in class or during office hours about these concerns.

Reflect on Your Math Attitude

     When many students see the words 'problem solving,' a sense of dread is experienced. How did you feel when you saw the title to this section?

If you felt just fine, or if you did not feel fine, why do you think that is so? How have your prior experiences influenced your feelings toward problem solving?

As you work through these problems, write any questions in the table in the *After Class* section under the title ***Getting to Work!***. Be sure to ask your instructor, tutor, or another student for help.

3.7 Problem Solving Using Systems of Three Equations: Pre-Class Prep

 Are You Ready?

Complete the following problems. These problems review some basic skills that are needed when solving application problems using systems of three equations. *All answers are found in the Pre-Class Prep Answer Section.*

1. Translate to mathematical symbols: 75 more than the number n.

2. Write an algebraic expression that represents the value of x coats, if each coat has a value of $150.

3. Write an algebraic expression that represents the value (in cents) of n nickels.

4. Solve the system $\begin{cases} x - y + 6z = 12 \\ x + 6y = -28 \\ 7y + z = -26 \end{cases}$ by substitution.

 Reading Time!

While **reading Section 3.7**, fill in the blanks choosing from the following words (some may be used more than once or not at all). *All answers are found in the Pre-Class Prep Answer Section.*

unknowns	matrices	problem-solving strategy
elimination	table	solve
$y = mx + b$	equation	three
two	curve fitting	check
$y = ax^2 + bx + c$	revenue	Cramer's rule

1 Assign Variables to Three Unknowns

1. The six step _____ can be extended to situations involving three unknowns.

2. To _____ the system of three equations in three variables, use either _____, substitution, _____, or _____.

3. Organize the facts of the problem in a _____.

4. Given the sale of x good hammers for $6 per hammer, the _____ received is $6x$.

5. If three variables represent the _____, _____ equations must be formed to solve the problem. To verify the result a three-part _____ must be performed.

2 Use Systems to Solve Curve-Fitting Problems

6. The process of determining an equation whose graph contains given points is called _____.

7. To determine the _____ of the parabola, substitute the x- and y-coordinates of three points that lie on the graph into the equation _____. Then _____ the system of _____ equations to find the values of a, b, and c.

✓ Getting Ready for Class

Briefly look through the section again. Answer the following by writing the concept or just the page number from the text.

Identify concepts/procedures that you feel confident about:

Identify concepts/procedures that look confusing or challenging:

Be sure to ask your instructor further questions if you are still having difficulty with a concept.

3.7 Problem Solving Using Systems of Three Equations: In-Class Notes

Terms, Definitions, and Main Ideas	Examples and Notes

Use notebook paper for additional notes

3.7 Problem Solving Using Systems of Three Equations: After Class

Important to Know

What is your homework assignment? Be sure to note it in your weekly schedule.

Section 3.7 Homework: _____ **Due:** _____

Getting to Work!

Complete your homework assignment. If you are unable to do a problem, write down the problem number and a question to help you remember what you would like to ask your instructor, your tutor, or another student.

Problem Number	Question? Where in the problem did you start to have difficulty or confusion?	Answered?

Often, you will have more questions than there is space provided here. If so, write them on notebook paper and be sure to talk to your instructor. You might ask in class or privately with the instructor.

Do You Really Know It?

Can you put into words the concepts that you learned in this section? Answer the below question from the *Writing* section in the *Study Set* in your text. Explain as if you were explaining to someone who has never taken this class before. Use notebook paper if you need more room.

Explain why the following problem does not give enough information to answer the question: The sum of three integers is 48. If the first integer is doubled, the sum is 60. Find the integers.

You Write the Test!

If you were writing the test for this section, what would you want a student to know? Write two test questions that you think might come from this material. Write questions of various difficulty; these questions can be original or chosen from the homework. Be sure to supply the answer also!

Write these questions at the end of this chapter under the section titled _My Practice Chapter 3 Test_ and the answer to each question under the section titled _My Practice Chapter 3 Test Answers_.

 ## Reflect on the Section

Look back at the _Pre-Class Prep_ section. Did the lecture explain topics that you thought were going to be challenging or confusing? _____

- Are there topics that you still have questions on from the reading or the lecture? If so, complete the following:
    I don't understand…

- Speak to your instructor in class or during office hours about these concerns.

 ## Reflect on Your Math Attitude

Think about the test-taking strategies suggested in How to Take a Math Test*. What strategies do you plan to use when you take your next test?

How does your attitude affect your performance on a test?

When you first get your test, write "_I can do this_" at the top – it does actually help!

Chapter 3 Activities

Your instructor may assign these activities to you to complete in class, or you may complete them on your own to solidify your understanding of chapter topics. The activities begin on the next page.

❖ **Student Activity:** *Choose Your Tactic*
  For each system of equations, determine your tactical plan: solving by substitution or solving by elimination.

❖ **Student Activity:** *Rowing to Freedom*
  Escape by completing elementary row operations.

❖ **Student Activity:** *D-lightful Determinants*
  Find the missing determinants, then shade to paint by determinants.

Student Activity
Choose Your Tactic

Directions: For each system of equations below, decide whether it will be easier to solve using substitution or elimination (choose a tactic). Then describe your tactical plan: For **substitution,** which variable in which equation will you solve for? For **elimination** – which variable will you eliminate and how?

	System of Equations	Tactic Substitution or Elimination	Tactical Plan Substitution: Which variable will you solve for? Elimination: Which variable will you eliminate?
1.	$\begin{cases} 5x+3y=-9 \\ y=2x+8 \end{cases}$		
2.	$\begin{cases} 2x+y=9 \\ 5x+3y=26 \end{cases}$		
3.	$\begin{cases} y=-x \\ 6x+6y=0 \end{cases}$		
4.	$\begin{cases} 4x+11y=7 \\ 4x+3y=-1 \end{cases}$		
5.	$\begin{cases} 16x-2y=16 \\ 4x=2y-8 \end{cases}$		
6.	$\begin{cases} 0.02x+0.01y=0.1 \\ -2x+3y=-18 \end{cases}$		
7.	$\begin{cases} -\frac{x}{5}-\frac{y}{3}=2 \\ -\frac{3x}{10}+\frac{2y}{10}=\frac{9}{10} \end{cases}$		
8.	$\begin{cases} x=12y-7 \\ 12y-x=12 \end{cases}$		

Hold the line men,
they're bringing in the -1 .

Cengage Student Workbook Activities, M. Andersen

Chapter 3 Systems of Equations

Student Activity
Rowing to Freedom

Directions: Begin with the matrix in the box marked "Start". Follow each of the given row operations and shade the box that contains your result for each one (each row operation is carried out on the previous answer). At the "pit stop" you'll get a new matrix to use.

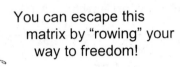

You can escape this matrix by "rowing" your way to freedom!

1. $-2R_2 \rightarrow R_2$

2. $R_2 + R_3 \rightarrow R_3$

3. $2R_1 + R_2 \rightarrow R_2$

4. $-\frac{3}{4}R_2 \rightarrow R_2$

5. $R_2 + R_3 \rightarrow R_3$

6. $-\frac{1}{3}R_2 \rightarrow R_2$ **(P.S.)**

7. $R_1 \leftrightarrow R_3$

8. $R_1 + (-2)R_2 \rightarrow R_2$

9. $R_2 + 2R_3 \rightarrow R_3$

10. $-1R_3 \rightarrow R_3$

11. $-\frac{1}{2}R_2 \rightarrow R_2$

12. $\frac{1}{4}R_1 \rightarrow R_1$

START $\begin{bmatrix} 1 & 1 & 1 & 0 \\ 1 & -1 & 1 & -6 \\ 2 & 1 & 3 & -2 \end{bmatrix}$	$\begin{bmatrix} 1 & 1 & 1 & 0 \\ -2 & 2 & -2 & 12 \\ 2 & 1 & 3 & -2 \end{bmatrix}$	$\begin{bmatrix} 1 & 1 & 1 & 0 \\ -2 & 2 & -2 & 12 \\ 0 & 3 & 1 & 10 \end{bmatrix}$	$\begin{bmatrix} 1 & 1 & 1 & 0 \\ 0 & 4 & 0 & 12 \\ 0 & 3 & 1 & 10 \end{bmatrix}$
$\begin{bmatrix} 1 & 1 & 1 & 0 \\ 1 & -1 & 1 & -6 \\ -4 & -2 & -6 & 4 \end{bmatrix}$	$\begin{bmatrix} -1 & 3 & -1 & 12 \\ -2 & 2 & -2 & 12 \\ 2 & 1 & 3 & -2 \end{bmatrix}$	$\begin{bmatrix} 1 & 1 & 1 & 0 \\ 0 & -1 & 0 & -3 \\ 0 & 0 & 1 & 1 \end{bmatrix}$	$\begin{bmatrix} 1 & 1 & 1 & 0 \\ 0 & -3 & 0 & -9 \\ 0 & 3 & 1 & 10 \end{bmatrix}$
$\begin{bmatrix} 4 & 2 & 1 & 5 \\ 2 & 2 & 2 & 2 \\ 0 & 1 & 1 & 1 \end{bmatrix}$	**PIT STOP** $\begin{bmatrix} 0 & 1 & 1 & 1 \\ 2 & 2 & 2 & 2 \\ 4 & 2 & 1 & 5 \end{bmatrix}$	$\begin{bmatrix} 1 & 1 & 1 & 0 \\ 0 & 1 & 0 & 3 \\ 0 & 0 & 1 & 1 \end{bmatrix}$	$\begin{bmatrix} 1 & 1 & 1 & 0 \\ 0 & -3 & 0 & -9 \\ 0 & 0 & 1 & 1 \end{bmatrix}$
$\begin{bmatrix} 4 & 2 & 1 & 5 \\ 0 & -2 & -3 & 1 \\ 0 & 1 & 1 & 1 \end{bmatrix}$	$\begin{bmatrix} 0 & 1 & 1 & 1 \\ 2 & 2 & 2 & 2 \\ 4 & 3 & 2 & 6 \end{bmatrix}$	$\begin{bmatrix} 1 & 1 & 1 & 0 \\ 0 & 1 & 0 & 0 \\ 0 & 0 & 1 & 1 \end{bmatrix}$	$\begin{bmatrix} 1 & 1 & 1 & 0 \\ 0 & 3 & 0 & 9 \\ 0 & 0 & 1 & 1 \end{bmatrix}$
$\begin{bmatrix} 4 & 2 & 1 & 5 \\ 0 & -2 & -3 & 1 \\ 0 & 0 & -1 & 3 \end{bmatrix}$	$\begin{bmatrix} 4 & 2 & 1 & 5 \\ 0 & 1 & \frac{3}{2} & -\frac{1}{2} \\ 0 & 0 & 1 & 0 \end{bmatrix}$	$\begin{bmatrix} 1 & \frac{1}{2} & \frac{1}{4} & \frac{5}{4} \\ 0 & 1 & \frac{3}{2} & 0 \\ 0 & 0 & 1 & 0 \end{bmatrix}$	$\begin{bmatrix} 1 & \frac{1}{2} & \frac{1}{4} & 0 \\ 0 & 1 & \frac{3}{2} & 0 \\ 0 & 0 & 1 & 0 \end{bmatrix}$
$\begin{bmatrix} 4 & 2 & 1 & 5 \\ 0 & -2 & -3 & 1 \\ 0 & 0 & 1 & -3 \end{bmatrix}$	$\begin{bmatrix} 4 & 2 & 1 & 5 \\ 0 & 1 & \frac{3}{2} & -\frac{1}{2} \\ 0 & 0 & 1 & -3 \end{bmatrix}$	$\begin{bmatrix} 1 & \frac{1}{2} & \frac{1}{4} & \frac{5}{4} \\ 0 & 1 & \frac{3}{2} & -\frac{1}{2} \\ 0 & 0 & 1 & -3 \end{bmatrix}$	**You've successfully "rowed" to freedom!**

Cengage Student Workbook Activities, M. Andersen

Student Activity
D-lightful Determinants

Directions: For each system of equations, find the desired determinants. You do **not** need to solve the system of equations. As you find the missing determinants, shade them in to paint by Determinants.

1. $\begin{cases} 2x+3y=-1 \\ y-x=8 \end{cases}$

   Find D_x and D_y.

2. $\begin{cases} x+2y+4z=4 \\ 3x+2y+6z=1 \\ x+5y+4z=1 \end{cases}$

   Find D_y and D_z.

3. $\begin{cases} 10x+12y+6z=2 \\ 5x+6y+3z=1 \\ x+3z=1 \end{cases}$

   Find D, D_x, D_y and D_z.

4. $\begin{cases} x+z=4 \\ 3y+3z=12 \\ x+2y=6 \end{cases}$

   Find D, D_x, D_y and D_z.

$\begin{vmatrix} 1 & 0 & 1 \\ 0 & 3 & 3 \\ 1 & 2 & 0 \end{vmatrix}$	$\begin{vmatrix} -1 & 3 \\ 8 & 1 \end{vmatrix}$	$\begin{vmatrix} 4 & 0 & 1 \\ 12 & 3 & 3 \\ 6 & 2 & 0 \end{vmatrix}$	$\begin{vmatrix} 2 & 3 \\ -1 & 1 \end{vmatrix}$	$\begin{vmatrix} 4 & 1 & 0 \\ 12 & 3 & 3 \\ 6 & 2 & 0 \end{vmatrix}$
$\begin{vmatrix} 2 & -1 \\ -1 & 8 \end{vmatrix}$	$\begin{vmatrix} 1 & 1 & 0 \\ 0 & 3 & 3 \\ 1 & 2 & 0 \end{vmatrix}$	$\begin{vmatrix} 1 & 2 & 4 \\ 3 & 2 & 6 \\ 1 & 5 & 4 \end{vmatrix}$	$\begin{vmatrix} 10 & 12 & 6 \\ 5 & 6 & 3 \\ 1 & 0 & 3 \end{vmatrix}$	$\begin{vmatrix} 10 & 12 & 6 \\ 5 & 6 & 3 \\ 1 & 3 & 0 \end{vmatrix}$
$\begin{vmatrix} 10 & 12 & 2 \\ 5 & 6 & 1 \\ 1 & 0 & 1 \end{vmatrix}$	$\begin{vmatrix} 10 & 12 & 2 \\ 5 & 6 & 1 \\ 1 & 3 & 1 \end{vmatrix}$	$\begin{vmatrix} 10 & 2 & 6 \\ 5 & 1 & 3 \\ 1 & 1 & 0 \end{vmatrix}$	$\begin{vmatrix} 10 & 2 & 6 \\ 5 & 1 & 3 \\ 1 & 1 & 1 \end{vmatrix}$	$\begin{vmatrix} 4 & 2 & 4 \\ 1 & 2 & 6 \\ 1 & 5 & 4 \end{vmatrix}$
$\begin{vmatrix} 1 & 0 & 4 \\ 0 & 3 & 12 \\ 1 & 2 & 6 \end{vmatrix}$	$\begin{vmatrix} 1 & 1 & 4 \\ 0 & 3 & 12 \\ 1 & 2 & 6 \end{vmatrix}$	$\begin{vmatrix} 1 & 3 \\ 8 & -1 \end{vmatrix}$	$\begin{vmatrix} 1 & 2 & 4 \\ 3 & 2 & 1 \\ 1 & 5 & 1 \end{vmatrix}$	$\begin{vmatrix} 2 & 3 \\ 1 & -1 \end{vmatrix}$
$\begin{vmatrix} 1 & 4 & 4 \\ 3 & 1 & 6 \\ 1 & 1 & 4 \end{vmatrix}$	$\begin{vmatrix} 2 & 12 & 6 \\ 1 & 6 & 3 \\ 1 & 0 & 3 \end{vmatrix}$	$\begin{vmatrix} 1 & 4 & 1 \\ 0 & 12 & 3 \\ 1 & 6 & 0 \end{vmatrix}$	$\begin{vmatrix} 2 & 12 & 6 \\ 1 & 6 & 3 \\ 1 & 3 & 0 \end{vmatrix}$	$\begin{vmatrix} 1 & 4 & 0 \\ 0 & 12 & 3 \\ 1 & 6 & 0 \end{vmatrix}$

Cengage Student Workbook Activities, M. Andersen

Chapter 3 Systems of Equations

Chapter 3 Test Skills Assessment

Pre-Test Preparation Work:

1. Re-read the objectives from each section.
2. Review the *Reading Time!* activity for each section.
3. Go over all your classroom notes, if something in your notes doesn't make sense to you, make a note and ask your teacher or a classmate.
4. Make additional notations to your work if your teacher states specific concepts to study in preparation for the chapter test.
5. Attend any study sessions held by your teacher or teaching assistant.
6. Practice additional problems.
7. Go over any missed problems in your homework sets.
8. Talk out concepts with your peers in small group study sessions.

List other preparations that you have found beneficial in preparing for a math test.

Additional Practice Suggestions

1. Use your book's review problems at the back of the chapter as a practice test.
2. Take *My Practice Chapter Test* and the text's Chapter Test. Time yourself and do not use your notebook or textbook.
3. Pace yourself as you work through these problems.
4. Read each question carefully, playing close attention to the instructions.
5. Check your work using the answers provided.
6. Rework any missed problems. Do not just "look them over" but actually rework the problem without looking at text or notes.

My Practice Chapter 3 Test

For each section, you had the opportunity to create two test questions under the section *You Write the Test!* Write each of those questions here. Include your answers under the heading *My Practice Chapter 3 Test Answers*. **Take the test without notes or your textbook.** If you do not get a question correct, review the text and/or your notes then take the test again. For further review, do the *Chapter 3 Test* in the text.

Section 3.1

1.

2.

Section 3.2

3.

4.

Section 3.3

5.

6.

Section 3.4

7.

8.

Section 3.5

9.

10.

Section 3.6

11.

12.

Section 3.7

13.

14.

My Practice Chapter 3 Test Answers

Section 3.1

1.

2.

Section 3.2

3.

4.

Section 3.3

5.

6.

Section 3.4

7.

8.

Section 3.5

9.

10.

Section 3.6

11.

12.

Section 3.7

13.

14.

Chapter 3 Systems of Equations

Chapter 4 Inequalities

Read the *Study Skills Workshop* found at the beginning of Chapter 4 in your textbook. **Complete** the activities below for this chapter's *Study Skills Workshop*.

Making Homework a Priority: *The only way to really learn algebra is to do the homework!*

When to Do Your Homework

Try to start your homework when the material is fresh in your mind. This allows you the time to ask questions before the homework is due. Also, it is best to break your homework sessions into 30-minute periods, allowing for short breaks in between.

Think about completing your algebra homework assignments. Complete the following.

✓ **What** is my homework assignment? Enter each section's homework in this notebook in the *After Class* section under the title **Important to Know**. You can also enter the due date.

✓ **When** I plan to complete my homework:

_____

✓ **Where** I plan to complete my homework:

_____

✓ **How** I plan to complete my homework:

_____

How to Begin Your Homework

Before starting your homework assignment, review your notes and the examples in your text. For each problem on your next homework assignment, find an example in the book that is similar. Write the example number next to the problem.

✓ I did this for homework assignment _____

✓ Did you find it helpful? If so, please describe how it was helpful. If not, why?

Getting Help with Your Homework

Have you ever had a question but when you got to class you forgot what the question was? Typically, you'll remember what the question was when you are in the middle of a quiz or test. Not so good! To prevent this, each time you have a question on a problem write it down. The question does not need to be wordy; it does need to be clear to you as to what you meant.

✓ In this notebook in the *After Class* section under the title *Getting to Work!*, is a table where you can write down the problem number and the question that you would like to ask your instructor, your tutor, or another student. Have your notebook open to *Getting to Work!* the next time you do your homework assignment!

4.1 Solving Linear Inequalities in One Variable: Pre-Class Prep

 Are You Ready?

Complete the following problems. These review some basic skills that are needed to solve inequalities. *All answers are found in the Pre-Class Prep Answer Section.*

1. Fill in the blank: The symbol > means "____ _____ ____."

2. Is $-11 > -10$ a true or false statement?

3. Graph each number in the set $\left\{ -3.8, -\frac{2}{5}, 0.6, \frac{9}{4} \right\}$ on a number line.

4. Express the fact that $6 > 0$ using an $<$ symbol.

 Reading Time!
(on the next page)

 Getting Ready for Class

Briefly look through the section again. Answer the following by writing the concept or just the page number from the text.

Identify concepts/procedures that you feel confident about:

Identify concepts/procedures that look confusing or challenging:

Be sure to ask your instructor further questions if you are still having difficulty with a concept.

Reading Time!

While **reading Section 4.1**, match the word or concept to its definition or description.
Answers are found in the Pre-Class Prep Answer Section.

1 Read and Interpret Inequality Symbols

_____ 1. Inequality

_____ 2. Example of a linear inequality in one variable

2 Graph Intervals and Use Interval and Set-Builder Notation

_____ 3. Interval

_____ 4. Parenthesis/Bracket

_____ 5. Interval notation

_____ 6. Positive infinity symbol ∞

_____ 7. Unbounded interval

3 Solve Linear Inequalities Using Properties of Inequality

_____ 8. Solution set

_____ 9. To solve a linear inequality

_____ 10. Equivalent inequalities

_____ 11. Addition Property of Inequality

_____ 12. Division Property of Inequality

_____ 13. Identities

_____ 14. Contradiction

4 Use Linear Inequalities to Solve Problems

_____ 15. "not more than" or "should exceed"

_____ 16. Form an inequality, then solve the inequality

A. inequality made true by any permissible replacement value for the variable

B. graph of a set of real numbers that is a portion of the number line

C. reverse the inequality when dividing both sides of an inequality by a negative

D. indicates that the interval continues without end to the right

E. set of all numbers that make the inequality true

F. a compact form of writing intervals, such as $(-3, \infty)$

G. indicates whether the endpoint is or is not part of the graph

H. statement indicating that two expressions are unequal and they contain one or more of the symbols $<, >, \leq, \geq, \neq$

I. steps 3 and 4 of the six-step problem solving strategy for problems involving inequalities

J. find all values of the variable that make the inequality true

K. inequalities that have the same solution set

L. $ax + b \geq c$, where a, b, and c are real numbers and $a \neq 0$

M. phrases that suggest that the problem involves an inequality

N. adding the same number to both sides of an inequality does not change its solutions

O. inequality made false for all replacement values for the variable

Q. an interval that extends forever in one direction

Chapter 4 Inequalities

4.1 Solving Linear Inequalities in One Variable: In-Class Notes

Terms, Definitions, and Main Ideas	Examples and Notes

Use notebook paper for additional notes

4.1 Solving Linear Inequalities in One Variable: After Class

Important to Know

What is your homework assignment? Be sure to note it in your weekly schedule.

Section 4.1 Homework: _____ **Due**: _____

Getting to Work!

Complete your homework assignment. If you are unable to do a problem, write down the problem number and a question to help you remember what you would like to ask your instructor, your tutor, or another student.

Problem Number	Question? Where in the problem did you start to have difficulty or confusion?	Answered?

Often, you will have more questions than there is space provided here. If so, write them on notebook paper and be sure to talk to your instructor. You might ask in class or privately with the instructor.

Do You Really Know It?

Can you put into words the concepts that you learned in this section? Answer the below question from the Writing section in the Study Set in your text. Explain as if you were explaining to someone who has never taken this class before. Use notebook paper if you need more room.

How are the methods for solving linear equations and linear inequalities similar? How are they different?

Chapter 4 Inequalities

© 2013 Cengage Learning. All Rights Reserved. May not be scanned, copied or duplicated, or posted to a publicly accessible website, in whole or in part.

You Write the Test!

If you were writing the test for this section, what would you want a student to know? Write test questions that you think might come from this material. Write questions of various difficulty; these questions can be original or chosen from the homework. Be sure to supply the answer also!

Write these questions at the end of this chapter under the section titled *My Practice Chapter 4 Test* and the answer to each question under the section titled *My Practice Chapter 4 Test Answers*.

Reflect on the Section

Look back at the *Pre-Class Prep* section. Did the lecture explain topics that you thought were going to be challenging or confusing? _____

- Are there topics that you still have questions on from the reading or the lecture? If so, complete the following:
    I don't understand…

- Speak to your instructor in class or during office hours about these concerns.

Reflect on Your Math Attitude

How did you feel when you saw the title to this section?

If you felt just fine, or if you did not feel fine, why do you think that is so? How have your prior experiences influenced your feelings toward this topic?

As you work through these problems, write any questions in the table in the *After Class* section under the title *Getting to Work!* Be sure to ask your instructor, tutor, or another student for help.

Are You Ready?

Complete the following problems. These review some basic skills that are needed to solve inequalities. *All answers are found in the Pre-Class Prep Answer Section.*

1. Let $A = \{-6, 1, 2, 3, 4\}$ and $B = \{0, 3, 4, 5, 6\}$. What numbers do sets A and B have in common?

2. Solve $6x + 8 + x \geq 5(x - 3) + 9$. Graph the solution set and write it in interval notation.

3. Consider the statement $x > 4$ and $x < 8$. Does the number 7 make both inequalities true?

4. Graph the solution set for $x > 2$ and the solution set for $x \leq -1$ on the same number line.

 Reading Time!

While **reading Section 4.2**, identify the word or concept being defined. Choose from the following words (some may be used more than once or not at all). *All answers are found in the Pre-Class Prep Answer Section.*

closed interval	intersection	all numbers that make both inequalities true
double inequalities	circle graphs	union of set A and set B
same number line	unbounded interval	$c < x$ and $x < d$
intersection of set A and set B	open intervals	interval notation or set-builder notation
half-open intervals	bounded interval	all numbers that make one or the other or both inequalities true
Venn diagrams	union	$x < c$ and $x > d$

1 Find the Intersection and the Union of Two Sets

1. Set of all elements that are common to set A and set B, written $A \cap B$: _____

2. Set of all elements that belong to set A or set B or both, written $A \cup B$: _____

3. Used to illustration the intersection and union of sets: _____

2 Solve Compound Inequalities Containing the Word And

4. Two inequalities joined with the word *and*: _____

5. Solution set of a compound inequality containing the word *and*: _____

6. Solution set of two inequalities when the endpoints are included: _____

7. Intervals which contain both endpoints; notation $[a,b]$: _____

8. Intervals which do not contain either endpoint; notation (a,b) : _____

9. Intervals which contain only one endpoint; notation $[a,b)$ or $(a,b]$: _____

3 Solve Double Linear Inequalities

10. Inequalities that contain exactly two inequality symbols: _____

11. The double inequality $c < x < d$ is equivalent to: _____

4 Solve Compound Inequalities Containing the Word Or

12. Solution set of a compound inequality containing the word *or*: _____

   Solving Compound Inequalities

13. Solve each inequality separately and graph their solution sets where: _____

14. For inequalities connected with word *and*, find what of the solution sets: _____

15. For inequalities connected with word *or*, find what of the solution sets: _____

16. Write the solution set of the compound inequality using this, then graph it on a number

   line: _____

Getting Ready for Class

Briefly look through the section again. Answer the following by writing the concept or just the page number from the text.

Identify concepts/procedures that you feel confident about:

Identify concepts/procedures that look confusing or challenging:

Be sure to ask your instructor further questions if you are still having difficulty with a concept.

Chapter 4 Inequalities

Terms, Definitions, and Main Ideas	Examples and Notes

Use notebook paper for additional notes

4.2 Solving Compound Inequalities: After Class

Important to Know

What is your homework assignment? Be sure to note it in your weekly schedule.

Section 4.2 Homework: _____ **Due**: _____

Getting to Work!

Complete your homework assignment. If you are unable to do a problem, write down the problem number and a question to help you remember what you would like to ask your instructor, your tutor, or another student.

Problem Number	Question? Where in the problem did you start to have difficulty or confusion?	Answered?

Often, you will have more questions than there is space provided here. If so, write them on notebook paper and be sure to talk to your instructor. You might ask in class or privately with the instructor.

Do You Really Know It?

Can you put into words the concepts that you learned in this section? Answer the below question from the Writing section in the **Study Set** in your text. Explain as if you were explaining to someone who has never taken this class before. Use notebook paper if you need more room.

What is incorrect about the double inequality $3 < -3x + 4 < -3$?

You Write the Test!

If you were writing the test for this section, what would you want a student to know? Write two test questions that you think might come from this material. Write questions of various difficulty and these questions can be original or chosen from the homework. Be sure to supply the answer also!

Write these questions at the end of this chapter under the section titled *My Practice Chapter 4 Test* and the answer to each question under the section titled *My Practice Chapter 4 Test Answers*.

Reflect on the Section

Look back at the *Pre-Class Prep* section. Did the lecture explain topics that you thought were going to be challenging or confusing? _____

- Are there topics that you still have questions on from the reading or the lecture? If so, complete the following:
  I don't understand…

- Speak to your instructor in class or during office hours about these concerns.

Reflect on Your Math Attitude

How has organizing when, where, and how you plan to do your homework affected your attitude toward completing your homework?

What could you do to make the experience of completing your homework more enjoyable? Yes, enjoyable!

Are You Ready?

Complete the following problems. These problems review some basic skills that are needed to solve absolute value equations and inequalities. *All answers are found in the Pre-Class Prep Answer Section.*

1. Find each absolute value.

   a. $\left|12\right|$ b. $\left|-7.5\right|$

2. Tell whether each statement is true or false.

   a. $\left|-3\right| \geq 2$ b. $\left|-26\right| < -27$

3. Solve $3x + 6 \leq -3$ or $3x + 6 \geq 9$. Graph the solution set and write it using interval notation.

4. Solve $-8 < 2x + 8 < 16$. Graph the solution set and write it using interval notation.

Reading Time!

While **reading Section 4.3**, identify the statements as True or False. *All answers are found in the Pre-Class Prep Answer Section.*

1 Solve Equations of the Form $\left|X\right| = k$

_____ 1. The solution of the absolute value equation $\left|x\right| = 5$ is all real numbers x whose distance from 0 on the number line is 5.

_____ 2. To solve $\left|X\right| = k$, solve the equivalent compound equation $X = k$ or $X = -k$.

_____ 3. The statement $X = k$ or $X = -k$ is called a compound equation because is consists of two separate equations.

_____ 4. To solve absolute value equations (or inequalities), isolate the absolute value expression on one side before writing the equivalent compound statement.

2 Solve Equations with Two Absolute Values

_____ 5. To solve $|X| = |Y|$, solve the compound equation $X = Y$ or $X = -Y$.

_____ 6. The opposite of the expression $3x + 25$ is $-3x + 25$.

3 Solve Inequalities of the Form $|X| < k$

_____ 7. The solution of the absolute value inequality $|x| < 5$ is all real numbers x whose distance from 0 on the number line is more than 5.

_____ 8. To solve $|X| < k$, solve the equivalent double inequality $-k < X < k$.

_____ 9. A tolerance range can be defined by an absolute value inequality.

4 Solve Inequalities of the form $|X| > k$

_____ 10. The solution of the absolute value inequality $|x| > 5$ is all real numbers x whose distance from 0 on the number line is greater than 5.

_____ 11. To solve $|X| \geq k$, solve the equivalent double inequality $X \leq -k$ or $X \geq k$.

_____ 12. The algebraic description corresponding to the geometric description, "$|x| < k$ means that x is less than k units from 0 on the number line," is "$|x| < k$ is equivalent to $X \leq -k$ or $X \geq k$."

Getting Ready for Class

Briefly look through the section again. Answer the following by writing the concept or just the page number from the text.

Identify concepts/procedures that you feel confident about:

Identify concepts/procedures that look confusing or challenging:

Be sure to ask your instructor further questions if you are still having difficulty with a concept.

4.3 Solving Absolute Value Equations and Inequalities: In-Class Notes

Terms, Definitions, and Main Ideas	Examples and Notes

Use notebook paper for additional notes

4.3 Solving Absolute Value Equations and Inequalities: After Class

Important to Know

What is your homework assignment? Be sure to note it in your weekly schedule.

Section 4.3 Homework: _____ **Due:** _____

Getting to Work!

Complete your homework assignment. If you are unable to do a problem, write down the problem number and a question to help you remember what you would like to ask your instructor, your tutor, or another student.

Problem Number	Question? Where in the problem did you start to have difficulty or confusion?	Answered?

Often, you will have more questions than there is space provided here. If so, write them on notebook paper and be sure to talk to your instructor. You might ask in class or privately with the instructor.

Do You Really Know It?

Can you put into words the concepts that you learned in this section? Answer the below question from the *Writing* section in the *Study Set* in your text. Explain as if you were explaining to someone who has never taken this class before. Use notebook paper if you need more room.

Explain why the equation $|x-4| = -5$ has no solution.

Chapter 4 Inequalities

You Write the Test!

If you were writing the test for this section, what would you want a student to know? Write two test questions that you think might come from this material. Write questions of various difficulty; these questions can be original or chosen from the homework. Be sure to supply the answer also!

Write these questions at the end of this chapter under the section titled *My Practice Chapter 4 Test* and the answer to each question under the section titled *My Practice Chapter 4 Test Answers*.

Reflect on the Section

Look back at the *Pre-Class Prep* section. Did the lecture explain topics that you thought were going to be challenging or confusing? _____

- Are there topics that you still have questions on from the reading or the lecture? If so, complete the following:
  I don't understand…

- Speak to your instructor in class or during office hours about these concerns.

Reflect on Your Math Attitude

Did this topic of solving absolute value inequalities make sense to you? If so or if not, how does that make you feel?

What can you do to maintain a positive attitude, or develop a positive attitude, about this new concept?

When you begin your homework, be sure to review your notes and the examples in your text.

4.4 Linear Inequalities in Two Variables: Pre-Class Prep

 Are You Ready?

Complete the following problems. These problems review some basic skills that are needed when graphing linear inequalities. *All answers are found in the Pre-Class Prep Answer Section.*

1. True or false: $4(-3)+10>-2$

2. True or false: $-6 \le -6$

3. Graph: $4x-5y=20$

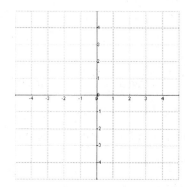

4. Determine whether each of the following points lies *above*, *below*, or *on* the line graphed in problem 3.

   a. $(2,-5)$

   b. $(0,0)$

   c. $(0,-4)$

 Reading Time!
   (on the next page)

 Getting Ready for Class

Briefly look through the section again. Answer the following by writing the concept or just the page number from the text.

Identify concepts/procedures that you feel confident about:

Identify concepts/procedures that look confusing or challenging:

Be sure to ask your instructor further questions if you are still having difficulty with a concept.

Chapter 4 Inequalities

Reading Time!

While **reading Section 4.4**, match the word or concept to its definition or description. *All answers are found in the Pre-Class Prep Answer Section.*

1 Graph Linear Inequalities

_____ 1. Examples of linear inequalities in two variables

_____ 2. Solution of a linear inequality in two variables

_____ 3. Graph of a linear inequality

_____ 4. Related equation to $y > 3x + 2$

_____ 5. Boundary line

_____ 6. Half-planes

_____ 7. Test point

_____ 8. Test-point Method

2 Graph Inequalities with a Boundary through the Origin

_____ 9. Test point for inequality with boundary through the origin

3 Graph Inequalities Having Horizontal and Vertical Boundary Lines

_____ 10. Equation of a horizontal boundary line

_____ 11. Equation of a vertical boundary line

4 Solve Applied Problems Involving Linear Inequalities in Two Variables

_____ 12. *At least, at most, should not exceed*

_____ 13. Quadrant I

A. $x = a$

B. $Ax + By < C$ or $Ax + By \geq C$

C. any point not on the boundary line; not (0,0)

D. point used to determine which half-plane contains the solution set

E. $y = b$

F. $y = 3x + 2$

G. separates the coordinate plane into two regions, may be solid or dashed

H. method used to graph linear inequalities in two variables

I. ordered pair whose coordinates satisfy the inequality

J. some phrases that indicate an inequality should be used

K. regions created by the boundary line dividing the coordinate plane

L. the quadrant in which to graph the inequality when only positive numbers can be solutions

M. graph of all ordered pairs whose coordinates satisfy the inequality

4.4 Linear Inequalities in Two Variables: In-Class Notes

Terms, Definitions, and Main Ideas	Examples and Notes

Use notebook paper for additional notes

Chapter 4 Inequalities

4.4 Linear Inequalities in Two Variables: After Class

 Important to Know

What is your homework assignment? Be sure to note it in your weekly schedule.

Section 4.4 Homework: _____   **Due**: _____

Getting to Work!

Complete your homework assignment. If you are unable to do a problem, write down the problem number and a question to help you remember what you would like to ask your instructor, your tutor, or another student.

Problem Number	Question? Where in the problem did you start to have difficulty or confusion?	Answered?

Often, you will have more questions than there is space provided here. If so, write them on notebook paper and be sure to talk to your instructor. You might ask in class or privately with the instructor.

Do You Really Know It?

Can you put into words the concepts that you learned in this section? Answer the below question from the *Writing* section in the *Study Set* in your text. Explain as if you were explaining to someone who has never taken this class before. Use notebook paper if you need more room.

Explain how to decide whether the boundary of the graph of a linear inequality should be drawn as a solid or a dashed line.

You Write the Test!

If you were writing the test for this section, what would you want a student to know? Write two test questions that you think might come from this material. Write questions of various difficulty; these questions can be original or chosen from the homework. Be sure to supply the answer also!

Write these questions at the end of this chapter under the section titled *My Practice Chapter 4 Test* and the answer to each question under the section titled *My Practice Chapter 4 Test Answers*.

Reflect on the Section

Look back at the *Pre-Class Prep* section. Did the lecture explain topics that you thought were going to be challenging or confusing? _____

- Are there topics that you still have questions on from the reading or the lecture? If so, complete the following:
  I don't understand…

- Speak to your instructor in class or during office hours about these concerns.

Reflect on Your Math Attitude

How do you feel about your homework? Describe what has or has not worked for you with respect to your homework.

Have any of the suggestions concerning homework presented in the Study Skills Workshop for this chapter helped you? If so, which ones?

Keeping your homework neat and organized will help you when you start preparing for quizzes or tests.

4.5 Systems of Linear Inequalities: Pre-Class Prep

Complete the following problems. These problems review some basic skills that are needed when graphing systems of linear inequalities. *All answers are found in the Pre-Class Prep Answer Section.*

1. Check to determine whether $(-3,5)$ is a solution of $2x - 3y \leq -25$.

2. Is the graph of the equation $x = 4$ on a rectangular coordinate system a vertical or horizontal line?

3. Graph: $x - 2y \geq 4$

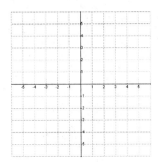

4. What inequality symbol is indicated by the phrase *at least*?

✓ **Reading Time!**

While **reading Section 4.5**, fill in the blanks choosing from the following words (some may be used more than once or not at all). *All answers are found in the Pre-Class Prep Answer Section.*

original	solution(s)	below	$>$
above	vertical	$\leq$	one
solution set	union	intersection	check
$<$	infinitely many	same	different
horizontal	$\geq$	systems of linear inequalities	

1 Solve Systems of Linear Inequalities

1. _____, such as $\begin{cases} y \leq -x+1 \\ 2x - y > 2 \end{cases}$, can be solved by graphing.

2. *Solving systems of Linear Inequalities*

   Step 1. Graph each inequality on the _____ rectangular coordinate system.

   Step 2. Use shading to highlight the _____ of the graphs. The points in this region are the _____ of the system.

   Step 3. As an informal _____, pick a point from the region and verify that its coordinates satisfy each inequality of the _____ system.

2 Graph Compound Inequalities

3. The _____ of double linear inequalities, such as $2 < x \leq 5$, can be graphed in the context of two variables.

4. The graph of $2 < x \leq 5$ will be two _____ lines, with shading in between the lines.

5. The graph of $y \geq -1$ or $y < -3$ on a rectangular coordinate system will be two _____ lines. The shading showing the solution set will be _____ the upper line and _____ the lower line.

3 Solve Problems Involving Systems of Linear Inequalities

6. Phrases such as *at most* and *should not surpass* can be represented by the inequality symbol _____.

7. Phrases such as *at least* and *cannot go below* can be represented by the inequality symbol _____.

Chapter 4 Inequalities

Getting Ready for Class

Briefly look through the section again. Answer the following by writing the concept or just the page number from the text.

Identify concepts/procedures that you feel confident about:

Identify concepts/procedures that look confusing or challenging:

Be sure to ask your instructor further questions if you are still having difficulty with a concept.

4.5 Systems of Linear Inequalities: In-Class Notes

<u>Terms, Definitions, and Main Ideas</u>	<u>Examples and Notes</u>

Use notebook paper for additional note

Chapter 4 Inequalities

4.5 Systems of Linear Inequalities: After Class

Important to Know

What is your homework assignment? Be sure to note it in your weekly schedule.

Section 4.5 Homework: _____ **Due**: _____

Getting to Work!

Complete your homework assignment. If you are unable to do a problem, write down the problem number and a question to help you remember what you would like to ask your instructor, your tutor, or another student.

Problem Number	Question? Where in the problem did you start to have difficulty or confusion?	Answered?

Often, you will have more questions than there is space provided here. If so, write them on notebook paper and be sure to talk to your instructor. You might ask in class or privately with the instructor.

Do You Really Know It?

Can you put into words the concepts that you learned in this section? Answer the below question from the *Writing* section in the *Study Set* in your text. Explain as if you were explaining to someone who has never taken this class before. Use notebook paper if you need more room.

Explain how a system of two linear inequalities might have no solution.

You Write the Test!

If you were writing the test for this section, what would you want a student to know? Write two test questions that you think might come from this material. Write questions of various difficulty; these questions can be original or chosen from the homework. Be sure to supply the answer also!

Write these questions at the end of this chapter under the section titled *My Practice Chapter 4 Test* **and the answer to each question under the section titled** *My Practice Chapter 4 Test Answers.*

Reflect on the Section

Look back at the *Pre-Class Prep* section. Did the lecture explain topics that you thought were going to be challenging or confusing? _____

- Are there topics that you still have questions on from the reading or the lecture? If so, complete the following:
    I don't understand…

- Speak to your instructor in class or during office hours about these concerns.

Reflect on Your Math Attitude

If you have followed the suggestions with homework that were provided in this chapter, how has that affected your attitude toward the course? If you did not do any of the suggestions, why didn't you do them?

What have you done with respect to homework that has most positively affected your attitude toward math and/or has made you more successful?

Chapter 4 Inequalities

Chapter 4 Activities

Your instructor may assign these activities to you to complete in class, or you may complete them on your own to solidify your understanding of chapter topics. The activities begin on the next page.

❖ **Student Activity:** *Graphing and Notation of Inequalities*

Represent an inequality in three ways: set notation, graph, and interval notation.

❖ **Student Activity:** *Following the Clues Back to the Inequality*

Given the graph of an inequality at each "crime-scene," determine what the inequality must have been to result in the graph.

❖ **Student Activity:** *Matching Up the Different Cases*

Categorize the absolute value equation or inequality as one of four cases or a special case.

Guided Learning Activity
Graphing and Notation of Inequalities

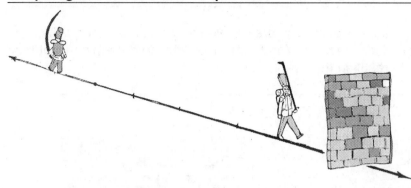

Think of interval notation as a way to tell someone how to draw the graph, from left to right, giving them only a "begin" value and an "end" value for each interval.

- Always give intervals from LEFT to RIGHT on the number line.
- Use $-\infty$ and ∞ to denote the "ends" of the number line (as shown above).
- Use (or) to denote an endpoint that is approached, but not included.
- Use [or] to denote an endpoint that is included.
- The parenthesis or bracket needs to open in the direction of the true part of the inequality.

Directions: Each inequality can be represented in three ways. One is given to you; fill in the missing ones.

	Set notation	Graph	Interval Notation
1.	$x > 2$	-5 -4 -3 -2 -1 0 1 2 3 4 5	
2.	$x \leq -3$	-5 -4 -3 -2 -1 0 1 2 3 4 5	
3.	$x \geq -1$	-5 -4 -3 -2 -1 0 1 2 3 4 5	
4.		-5 -4 -3 -2 -1 0 1 2 3 4 5	
5.		-5 -4 -3 -2 -1 0 1 2 3 4 5	
6.	$-2 < x \leq 3$	-5 -4 -3 -2 -1 0 1 2 3 4 5	
7.		-5 -4 -3 -2 -1 0 1 2 3 4 5	
8.		-5 -4 -3 -2 -1 0 1 2 3 4 5	$(-\infty, 4)$
9.		-5 -4 -3 -2 -1 0 1 2 3 4 5	$(-2, 0]$
10.		-5 -4 -3 -2 -1 0 1 2 3 4 5	$(-\infty, \infty)$

Cengage Student Workbook Activities, M. Andersen

Student Activity
Following the Clues Back to the Inequality

Directions: In each "crime-scene" below, you are shown the graph of an inequality. Use your mathematical powers of reasoning (and detective skills) to determine what the inequality must have been to result in this graph.

1.

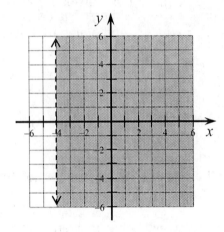

3.

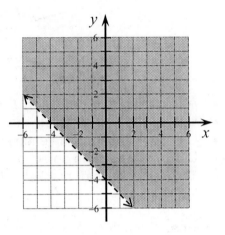

2.

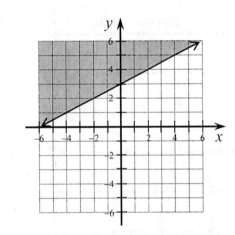

looks like a trail of clues...

(1,2) (3,7) (2,5) (6,4) (9,3)

Student Activity
Matching Up the Different Cases

Directions: Begin by isolating the absolute value in each of the equations and inequalities below. Then categorize the problem as one of the four cases or a special case. The first one has been done for you.

Let k be a **positive** constant and let X, X_1, and X_2 be mathematical expressions.

Case 1:	Case 2:	Case 3:	Case 4:	Special Case:
$\lvert X \rvert = k$	$\lvert X_1 \rvert = \lvert X_2 \rvert$	$\lvert X \rvert < k$ $\lvert X \rvert \leq k$	$\lvert X \rvert > k$ $\lvert X \rvert \geq k$	If the constant is negative or zero.

$2\lvert x-3 \rvert < 4$ $\lvert x-3 \rvert < 2$ **CASE 3**	$\lvert x-3 \rvert - 4 \geq 2$	$\lvert 2a-3 \rvert + 5 > 3$	$\dfrac{\lvert x \rvert}{3} - 2 = 4$
$\lvert x+4 \rvert = \lvert 2x-2 \rvert$	$1 = \lvert 2t-4 \rvert + 4$	$5 - \lvert 2x \rvert < 2$	$4\lvert x+3 \rvert - 3 \leq 2$
$-2\lvert 4y \rvert < 10$	$\dfrac{\lvert 3x-4 \rvert}{-6} \leq -2$	$1 > \lvert 3-b \rvert - 6$	$\lvert 2x \rvert = \lvert x+3 \rvert$
$5 - \lvert x+4 \rvert = 2$	$6 - 2\lvert x+4 \rvert \leq 0$	$\lvert x-4 \rvert + 5 = 5$	$3 < \dfrac{\lvert d+12 \rvert}{5}$

Cengage Student Workbook Activities, M. Andersen

Chapter 4 Inequalities

Chapter 4 Test Skills Assessment

Pre-Test Preparation Work:

1. Re-read the objectives from each section.
2. Review the *Reading Time!* activity for each section.
3. Go over all your classroom notes, if something in your notes doesn't make sense to you, make a note and ask your teacher or a classmate.
4. Make additional notations to your work if your teacher states specific concepts to study in preparation for the chapter test.
5. Attend any study sessions held by your teacher or teaching assistant.
6. Practice additional problems.
7. Go over any missed problems in your homework sets.
8. Talk out concepts with your peers in small group study sessions.

List other preparations that you have found beneficial in preparing for a math test.

Additional Practice Suggestions

1. Use your book's review problems at the back of the chapter as a practice test.
2. Take *My Practice Chapter Test* and the text's *Chapter Test*. Time yourself and do not use your notebook or textbook.
3. Pace yourself as you work through these problems.
4. Read each question carefully, playing close attention to the instructions.
5. Check your work using the answers provided.
6. Rework any missed problems. Do not just "look them over" but actually rework the problem without looking at text or notes.

My Practice Chapter 4 Test

For each section, you had the opportunity to create two test questions under the section *You Write the Test!* Write each of those questions here. Include your answers under the heading *My Practice Chapter 4 Test Answers*. **Take the test without notes or your textbook.** If you do not get a question correct, review the text and/or your notes then take the test again. For further review, do the *Chapter 4 Test* in the text.

Section 4.1

1.

2.

Section 4.2

3.

4.

Section 4.3

5.

6.

Section 4.4

7.

8.

Section 4.5

9.

10.

Chapter 4 Inequalities

My Practice Chapter 4 Test Answers

Section 4.1

1.

2.

Section 4.2

3.

4.

Section 4.3

5.

6.

Section 4.4

7.

8.

Section 4.5

9.

10.

Chapter 5 Exponents, Polynomials, and Polynomial Functions

Read the *Study Skills Workshop* found at the beginning of Chapter 5 in your textbook. **Complete** the activities below for this chapter's *Study Skills Workshop*.

Attending Class Regularly: Missing class or being late can negatively affect your grade.

Arrive On Time, Or a Little Early

Arriving on time or even a little early is a good way to begin a class. When do you typically arrive to class? Check your response.

_____ On time   _____ Early   _____ Late   _____ Other _____

✓ When you arrive, which of the following do you do?
   _____ visit with friends

   _____ get out your note-taking materials and homework

   _____ try to catch a quick nap

   _____ identify any questions that you plan to ask your instructor once class starts

Being prepared as the instructor begins the class will help you to have a better classroom experience.

If You Must Miss a Class

Planning ahead will often help you to avoid missing a class.

✓ List some reasons why you might miss class.

_____

_____

What can you do ahead of time so that these situations won't cause you to be tardy or absent?

_____

_____

Study the Material You Missed

There are online resources that are available with this textbook, such as video examples and problem-specific tutorials.
✓ Watch one section from the videotape series that accompanies this book. Take notes as you watch the explanations.

✓ Are You Ready?

Complete the following problems. These review some basic skills that are needed when working with exponents. *All answers are found in the Pre-Class Prep Answer Section.*

1. Evaluate:
   a. $6 + 6 + 6$

   b. $6 \cdot 6 \cdot 6$

2. Evaluate:
   a. 2^5

   b. $2 \cdot 5$

3. Write the expression $x \cdot x \cdot x \cdot x \cdot x \cdot x$ in an equivalent form using an exponent.

4. Translate to mathematical symbols:
   a. y squared   b. 7 cubed

5. What is the reciprocal of 8?

6. Fill in the blank: $9m^4$ means $9 \underline{\quad} m^4$.

Reading Time!

While **reading Section 5.1**, identify the word or concept being defined. Choose from the following words (some may be used more than once or not at all). *All answers are found in the Pre-Class Prep Answer Section.*

power of a product rule	exponential expressions	negative exponents appearing in fractions
exponent	quotient rule (for exponents)	exponents of 0 and 1
power rule (for exponents)	power of a quotient rule	product rule (for exponents)
natural-number exponent	x^n	1
indeterminate form	negative exponent(s)	base
$x^m \cdot x^n = x^{m+n}$	rule for negative exponents and reciprocals	
$x^0 = 1$	$\dfrac{y^n}{x^m}$	

1 Identify Bases and Exponents

1. Provides a way to write a product of repeated factors in compact form: _____

2. The factor that is raised to a power: _____

3. Tells how many times its base is to be used as a factor: _____

4. Expressions of the form x^n: _____

5. Exponent indicating that the base is used as a factor 1 time: _____

2 Use the Product and Power Rules for Exponents

6. For any number x and any natural numbers m and n, $x^m \cdot x^n = x^{m+n}$: _____

7. For any number x and any natural numbers m and n, $\left(x^m\right)^n = x^{mn}$: _____

8. For any numbers x and y, and any natural number n, $(xy)^n = x^n y^n$: _____

9. For any numbers x and y, and any natural number n, $\left(\dfrac{x}{y}\right)^n = \dfrac{x^n}{y^n}$: _____

3 Use the Zero and Negative-Integer Exponent Rules

10. Any nonzero base raised to the 0 power is 1: _____

11. 0^0 is called this because it is undefined: _____

12. For any nonzero real number x and any integer n, $x^{-n} = \dfrac{1}{x^n}$: _____

13. The reciprocal of x^{-n} : _____

14. Simplification of $\dfrac{1}{x^{-n}}$: _____

15. Changing from negative to positive exponents, $\dfrac{x^{-m}}{y^{-n}} = ?$: _____

4 Use the Quotient Rule for Exponents

16. For any nonzero number x and any integers m and n, $\dfrac{x^m}{x^n} = x^{m-n}$: _____

5 Simplify Quotients Raised to Negative Powers

17. For any nonzero real numbers x and y, and any integer n, $\left(\dfrac{x}{y}\right)^{-n} = \left(\dfrac{y}{x}\right)^n$: _____

18. Summary of Exponent Rules

   i. $x^m \cdot x^n = x^{m+n}$:_____

   ii. $\left(x^m\right)^n = x^{mn}$:_____

   iii. $(xy)^n = x^n y^n$:_____

   iv. $\dfrac{x^m}{x^n} = x^{m-n}$:_____

   v. $\left(\dfrac{x}{y}\right)^n = \dfrac{x^n}{y^n}$:_____

   vi. $x^0 = 1$ and $x^1 = x$:_____

   vii. $x^{-n} = \dfrac{1}{x^n}$:_____

   viii. $\dfrac{1}{x^{-n}} = x^n, \dfrac{x^{-m}}{y^{-n}} = \dfrac{y^n}{x^m}, \left(\dfrac{x}{y}\right)^{-n} = \left(\dfrac{y}{x}\right)^n$:_____

Getting Ready for Class

Briefly look through the section again. Answer the following by writing the concept or just the page number from the text.

Identify concepts/procedures that you feel confident about:

Identify concepts/procedures that look confusing or challenging:

Be sure to ask your instructor further questions if you are still having difficulty with a concept.

5.1 Exponents: In-Class Notes

Terms, Definitions, and Main Ideas

Examples and Notes

Use notebook paper for additional notes

5.1 Exponents: After Class

Important to Know

What is your homework assignment? Be sure to note it in your weekly schedule.

Section 5.1 Homework: _____ **Due**: _____

Getting to Work!

Complete your homework assignment. If you are unable to do a problem, write down the problem number and a question to help you remember what you would like to ask your instructor, your tutor, or another student.

Problem Number	Question? Where in the problem did you start to have difficulty or confusion?	Answered?

Often, you will have more questions than there is space provided here. If so, write them on notebook paper and be sure to talk to your instructor. You might ask in class or privately with the instructor.

Do You Really Know It?

Can you put into words the concepts that you learned in this section? Answer the below question from the *Writing* section in the *Study Set* in your text. Explain as if you were explaining to someone who has never taken this class before. Use notebook paper if you need more room.

Explain the error in the following solution.

Write $-8ab^{-3}$ *using positive exponents only.*

$$\cancel{-8ab^{-3} = \frac{a}{8b^3}}$$

You Write the Test!

If you were writing the test for this section, what would you want a student to know? Write two test questions that you think might come from this material. Write questions of various difficulty; these questions can be original or chosen from the homework. Be sure to supply the answer also!

Write these questions at the end of this chapter under the section titled *My Practice Chapter 5 Test* and the answer to each question under the section titled *My Practice Chapter 5 Test Answers*.

Reflect on the Section

Look back at the *Pre-Class Prep* section. Did the lecture explain topics that you thought were going to be challenging or confusing? _____

- Are there topics that you still have questions on from the reading or the lecture? If so, complete the following:
    I don't understand…

- Speak to your instructor in class or during office hours about these concerns.

Reflect on Your Math Attitude

As the middle of the term arrives, at what level is your enthusiasm for this class?

In what way has your enthusiasm level affected the amount of time that you spend either attending this class or working on homework?

Perseverance is often necessary whenever one's enthusiasm is waning. Think of the goal that you are accomplishing by completing this class well. Seek help from your instructor, tutor, or another student as needed.

5.2 Scientific Notation: Pre-Class Prep

 Are You Ready?

Complete the following problems. These problems review some basic skills that are needed when working with scientific notation. *All answers are found in the Pre-Class Prep Answer Section.*

1. Evaluate: 10^4

2. Multiply: $1,000,000 \cdot 9.63$

3. Evaluate: 10^{-3}

4. Multiply: $0.01 \cdot 4.31$

Reading Time!

While **reading Section 5.2**, identify the statements as True or False. *All answers are found in the Pre-Class Prep Answer Section.*

1 Write Numbers in Scientific Notation

_____ 1. A positive number is written in scientific notation when it is written in the form $N \times 10^n$, where $1 \leq N \leq 10$ and n is an integer.

_____ 2. Scientific notation provides a compact way of writing very large or very small numbers.

_____ 3. Standard notation is also called decimal notation.

_____ 4. The number 10^{-4} is a power of 10.

_____ 5. First determine n, then N to write a number in scientific notation.

_____ 6. The number 432×10^5 is not in scientific notation.

2 Convert from Scientific Notation to Standard Notation

_____ 7. To convert from scientific to standard notation, if the exponent is negative, move the decimal point the same number of places to the right as the absolute value of the exponent.

_____ 8. The form of scientific numbers for real numbers between 0 and 1: $\square \times 10^{positive\,integer}$.

_____ 9. The form of scientific numbers for real numbers greater than or equal to 10:

$\square \times 10^{\text{positive integer}}$.

3 Perform Calculations with Scientific Notation

_____ 10. The product of two numbers written in scientific notation is $(a \cdot b) \times 10^{m-n}$.

_____ 11. The quotient of two numbers written in scientific notation is $\left(\dfrac{a}{b}\right) \times 10^{m-n}$.

Getting Ready for Class

Briefly look through the section again. Answer the following by writing the concept or just the page number from the text.

Identify concepts/procedures that you feel confident about:

Identify concepts/procedures that look confusing or challenging:

Be sure to ask your instructor further questions if you are still having difficulty with a concept.

5.2 Scientific Notation: In-Class Notes

Terms, Definitions, and Main Ideas	Examples and Notes

Use notebook paper for additional notes

5.2 Scientific Notation: After Class

Important to Know

What is your homework assignment? Be sure to note it in your weekly schedule.

Section 5.2 Homework: _____ **Due**: _____

Getting to Work!

Complete your homework assignment. If you are unable to do a problem, write down the problem number and a question to help you remember what you would like to ask your instructor, your tutor, or another student.

Problem Number	Question? Where in the problem did you start to have difficulty or confusion?	Answered?

Often, you will have more questions than there is space provided here. If so, write them on notebook paper and be sure to talk to your instructor. You might ask in class or privately with the instructor.

Do You Really Know It?

Can you put into words the concepts that you learned in this section? Answer the below question from the *Writing* section in the *Study Set* in your text. Explain as if you were explaining to someone who has never taken this class before. Use notebook paper if you need more room.

Explain why 9.99×10^n represents a number less than 1 but greater than 0 if n is a negative integer.

You Write the Test!

If you were writing the test for this section, what would you want a student to know? Write two test questions that you think might come from this material. Write questions of various difficulty; these questions can be original or chosen from the homework. Be sure to supply the answer also!

Write these questions at the end of this chapter under the section titled *My Practice Chapter 5 Test* and the answer to each question under the section titled *My Practice Chapter 5 Test Answers*.

Reflect on the Section

Look back at the *Pre-Class Prep* section. Did the lecture explain topics that you thought were going to be challenging or confusing? _____

- Are there topics that you still have questions on from the reading or the lecture? If so, complete the following:
    I don't understand…

- Speak to your instructor in class or during office hours about these concerns.

Reflect on Your Math Attitude

How important is it to you that you attend class? Circle one.

Not important   Slightly important   Very important

How will your attitude toward attendance affect your grade in this course?

Make sure that your attitudes and goals align!

Are You Ready?

Complete the following problems. These problems review some basic skills that are needed when working with polynomials and polynomial functions. *All answers are found in the Pre-Class Prep Answer Section.*

1. a. How many terms does the expression $4x^2 - 8x + 1$ have?

   b. What is the coefficient of the second term?

2. Fill in the blanks: The equation $f(x) = x - 3$ defines a _____, because to each value of x in the _____, there corresponds exactly one value of $f(x)$ in the range.

3. Let $f(x) = 7x + 11$. Find $f(-3)$.   4. Combine like terms: $7.4a^3 + 2.8a^3$

5. Combine like terms: $6t^2u - \left(-11t^2u\right)$ 6. Simplify: $-\left(3b^2 - b + 24\right)$

 Reading Time!

While **reading Section 5.3**, fill in the blanks choosing from the following words (some may be used more than once). *All answers are found in the Pre-Class Prep Answer Section.*

evaluate	output	parabolas	exponent(s)	monomial
descending	degree of a term	1 (one)	opposites	like terms
combine like terms	polynomial	ascending	change the signs	linear
−1	quadratic	0 (zero)	coefficient	binomial
trinomial	leading	no defined degree	subtract	add
input	cubic	continuous	same	smooth
polynomial in three variables		sum	polynomial in one variable	
polynomial in two variables		degree of the polynomial		

1 Define and Classify Polynomials

1. A single term or a sum of terms in which all variables have whole-number _____ and no variable appears in a denominator is a _____.

2. An example of a _____ is $5x + 3$.

3. The polynomial $4n^2 - 6n - 8$ is written in _____ powers of n. The _____ term is $4n^2$ and 4 is the leading _____.

4. A polynomial can have more than _____ variable. For example, $p^3 + 3p^2q + 3pq^2 + q^3$ is a _____ and $-\dfrac{5}{2}rs^2t^4$ is a _____.

5. The polynomial $p^3 + 3p^2q + 3pq^2 + q^3$ is written in _____ powers of q.

6. A polynomial with 1 term is called a _____, a polynomial with 2 terms is called a _____, and a polynomial with three terms is called a _____.

7. The _____ of a polynomial in one variable is the value of the _____ on the variable. If a polynomial is in more than one variable, the degree of a term is the _____ of the exponents on the variables in that term. The degree of a nonzero constant is _____. The constant 0 has _____.

8. The highest degree of any term of the polynomial is the _____.

2 Evaluate Polynomials Functions

9. _____ equations are defined by equations of the form $f(x) = mx + b$. A second-degree polynomial function is called a _____ function and a third-degree polynomial function is called a _____ function.

10. To _____ a polynomial at a specific value, replace the variable in the equation with that value, called the _____. Simplify the result to find the _____.

3 Find Function Values and the Domain and Range of Polynomial functions Graphically

11. Graphs of polynomial functions of degree _____ are straight lines. The graphs of polynomial functions of degree 2 are _____. All polynomial function graphs are always _____ and _____.

4 Simplify Polynomials by Combining Like Terms

12. Terms with the same variables with the same exponents are called _____.

13. To _____ combine their coefficients and keep the same variables with the _____ exponents.

5 Add Polynomials

14. To _____ polynomials, drop the parentheses and combine their like terms.

6 Subtract Polynomials

15. To _____ two polynomials, _____ of the terms of the polynomial being subtracted, drop the parentheses, and combine like terms.

16. Just like real numbers, polynomials have _____, which are found by multiplying each term by _____.

✓ Getting Ready for Class

Briefly look through the section again. Answer the following by writing the concept or just the page number from the text.

Identify concepts/procedures that you feel confident about:

Identify concepts/procedures that look confusing or challenging:

Be sure to ask your instructor further questions if you are still having difficulty with a concept.

5.3 Polynomials and Polynomial Functions: In-Class Notes

Terms, Definitions, and Main Ideas	Examples and Notes

Use notebook paper for additional notes

Important to Know

What is your homework assignment? Be sure to note it in your weekly schedule.

Section 5.3 Homework: _____ **Due**: _____

Getting to Work!

Complete your homework assignment. If you are unable to do a problem, write down the problem number and a question to help you remember what you would like to ask your instructor, your tutor, or another student.

Problem Number	Question? Where in the problem did you start to have difficulty or confusion?	Answered?

Often, you will have more questions than there is space provided here. If so, write them on notebook paper and be sure to talk to your instructor. You might ask in class or privately with the instructor.

Do You Really Know It?

Can you put into words the concepts that you learned in this section? Answer the below question from the *Writing* section in the *Study Set* in your text. Explain as if you were explaining to someone who has never taken this class before. Use notebook paper if you need more room.

Explain why $f(x) = \dfrac{1}{x+1}$ *is not a polynomial function.*

You Write the Test!

If you were writing the test for this section, what would you want a student to know? Write two test questions that you think might come from this material. Write questions of various difficulty; these questions can be original or chosen from the homework. Be sure to supply the answer also!

Write these questions at the end of this chapter under the section titled *My Practice Chapter 5 Test* and the answer to each question under the section titled *My Practice Chapter 5 Test Answers*.

Reflect on the Section

Look back at the *Pre-Class Prep* section. Did the lecture explain topics that you thought were going to be challenging or confusing? _____

- Are there topics that you still have questions on from the reading or the lecture? If so, complete the following:
  I don't understand…

- Speak to your instructor in class or during office hours about these concerns.

Reflect on Your Math Attitude

Have you found the online resources that are available with this textbook? If so, what features have you like the best? If not, please check them out.

Has using the online resources affected your confidence in your ability to be successful?

Being able to see concepts presented in a variety of ways will help you to learn.

Are You Ready?

Complete the following problems. These problems review some basic skills that are needed when multiplying polynomials. *All answers are found in the Pre-Class Prep Answer Section.*

1. Multiply: $9 \cdot 10a$

2. Multiply: $5(2x-3)$

3. Multiply: $(4y+8)7$

4. Multiply: $x^8 \cdot x^6$

5. Multiply: $-2(3)(5)$

6. Simplify by combining like terms:
$$9x^2 + 6x - 3x + 2$$

 Reading Time!

(on the next page)

 Getting Ready for Class

Briefly look through the section again. Answer the following by writing the concept or just the page number from the text.

Identify concepts/procedures that you feel confident about:

Identify concepts/procedures that look confusing or challenging:

Be sure to ask your instructor further questions if you are still having difficulty with a concept.

Reading Time!

While **reading Section 5.4**, match the word or concept to its definition or description. Not all choices are used. *All answers are found in the Pre-Class Prep Answer Section.*

1 Multiply Monomials

_____ 1. Multiplying monomials

2 Multiply a Polynomial by a Monomial

_____ 2. Multiplying a polynomial by a monomial

3 Multiply Two Binomials

_____ 3. Multiplying two binomials

_____ 4. FOIL Method

4 Multiply Any Two Polynomials

_____ 5. Multiplying two polynomials

_____ 6. Vertical form

5 Multiply Three Polynomials

_____ 7. Multiplying three polynomials

6 Find Special Products

_____ 8. Square of a sum: $(A+B)^2 = ?$

_____ 9. Square of a difference: $(A-B)^2 = ?$

_____ 10. Product of the sum and difference of two terms: $(A+B)(A-B) = ?$

_____ 11. The result of squaring a binomial

_____ 12. Special-product rules

7 Use Multiplication to Simplify Expressions

_____ 13. To simplify algebraic expressions that involve the multiplication of polynomials

A. multiply any two polynomials, then multiply that result by the third polynomial

B. $A^2 - B^2$

C. multiply each term of one binomial by each term of the other binomial, then combine like terms

D. multiply numerical factors and the variable factors

E. $A^2 + 2AB + B^2$

F. shortcut method to multiply two binomials by multiplying first terms, outer terms, inner terms, and last terms, then combine like terms

G. perfect-square trinomial

H. multiply each term of one polynomial by each term of the other, then combine like terms

I. multiply each term of the polynomial by the monomial

J. multiply using an approach similar to that used to multiply whole numbers

K. $A^2 - 2AB + B^2$

L. use special-product rules and FOIL method. Then combine like terms.

M. rules for certain products that occur often

Chapter 5 Exponents, Polynomials, and Polynomial Functions

5.4 Multiplying Polynomials: In-Class Notes

Terms, Definitions, and Main Ideas	Examples and Notes

Use notebook paper for additional notes

5.4 Multiplying Polynomials: After Class

 Important to Know

What is your homework assignment? Be sure to note it in your weekly schedule.

Section 5.4 Homework: _____ **Due**: _____

 Getting to Work!

Complete your homework assignment. If you are unable to do a problem, write down the problem number and a question to help you remember what you would like to ask your instructor, your tutor, or another student.

Problem Number	Question? Where in the problem did you start to have difficulty or confusion?	Answered?

Often, you will have more questions than there is space provided here. If so, write them on notebook paper and be sure to talk to your instructor. You might ask in class or privately with the instructor.

 Do You Really Know It?

Can you put into words the concepts that you learned in this section? Answer the below question from the *Writing* section in the *Study Set* in your text. Explain as if you were explaining to someone who has never taken this class before. Use notebook paper if you need more room.

Explain how you would multiply two trinomials.

You Write the Test!

If you were writing the test for this section, what would you want a student to know? Write two test questions that you think might come from this material. Write questions of various difficulty; these questions can be original or chosen from the homework. Be sure to supply the answer also!

Write these questions at the end of this chapter under the section titled *My Practice Chapter 5 Test* and the answer to each question under the section titled *My Practice Chapter 5 Test Answers*.

Reflect on the Section

Look back at the *Pre-Class Prep* section. Did the lecture explain topics that you thought were going to be challenging or confusing? _____

- Are there topics that you still have questions on from the reading or the lecture? If so, complete the following:
    I don't understand…

- Speak to your instructor in class or during office hours about these concerns.

Reflect on Your Math Attitude

How did you feel when you learned that multiplying polynomials can be done in a vertical format just as with multiplying whole numbers?

Think of other instances when what seemed new was actually something old, just in a different form? How does that knowledge affect your attitude toward the new topic?

Look for similarities with material previously learned as you are learning new concepts.

5.5 The Greatest Common Factor and Factoring by Grouping: Pre-Class Prep

Are You Ready?

Complete the following problems. These review some basic skills that are needed when factoring expressions. *All answers are found in the Pre-Class Prep Answer Section.*

1. Find the prime factorization of 108.

2. Write $t \cdot t \cdot t \cdot t \cdot t$ in an equivalent form using exponents.

3. Simplify: $-(h-7)$

4. How many terms does the expression $3n^3 + n^2 + 4n + 8$ have?

5. Multiply: $6b(b^3 + 2b)$

6. Multiply and simplify: $m(m+2) - 5(m+2)$

Reading Time!

While **reading Section 5.5**, fill in the blanks choosing from the following words (some may be used more than once or not at all). *All answers are found in the Pre-Class Prep Answer Section.*

GCF	0	first	prime	common	factoring
regroup	coefficient	factor(s)	1	factoring out	isolate
product	variable	multiplication	greatest	opposite	last
grouping	common factor	irreducible	group	completely	formula

1 Find the Greatest Common Factor of a List of Terms

1. The largest common factor of a list of integers is called the _____ common _____ (_____).

2. To find the GCF of a list of terms, first write each _____ as a product of _____ numbers, then identify the numerical and _____ factors _____ to each term. Multiply the common numerical and variable _____.

3. If there are no common factors, the GCF is _____.

2 Factor Out the Greatest Common Factor

4. To write $6x^5 + 8x^3$ as $2x^3(3x^2 + 4)$, use the process _____ the greatest common factor.

5. A polynomial that cannot be factored is called a _____ polynomial or an _____ polynomial.

6. To factor a polynomial whose leading coefficient is negative, factor out the _____ of the GCF.

3 Factor by Grouping

7. The method of arranging terms in convenient groups is called _____ by _____.

8. *Factoring by Grouping*

   Step 1: _____ the terms of the polynomial so that each group has a common _____.

   Step 2: Factor out the _____ factor from each _____.

   Step 3: Factor out the resulting _____. If there is no common factor, _____ the terms of the polynomial and repeat steps 2 and 3.

9. By the _____ property of _____, the number _____ is a factor of every term.

10. When each _____ of a factored expression is prime, the polynomial is factored _____.

11. When factoring a polynomial, always look for a common factor _____.

4 Use Factoring to Solve Formulas for a Specified Variable

12. To solve a _____ for a specified _____ means to _____ that variable on one side of the equation, with all other variables and constants on the _____ side.

✓ Getting Ready for Class

Briefly look through the section again. Answer the following by writing the concept or just the page number from the text.

Identify concepts/procedures that you feel confident about:

Identify concepts/procedures that look confusing or challenging:

Be sure to ask your instructor further questions if you are still having difficulty with a concept.

Terms, Definitions, and Main Ideas	Examples and Notes

Use notebook paper for additional notes

5.5 The Greatest Common Factor and Factoring by Grouping: After Class

Important to Know

What is your homework assignment? Be sure to note it in your weekly schedule.

Section 5.5 Homework: _____ **Due:** _____

Getting to Work!

Complete your homework assignment. If you are unable to do a problem, write down the problem number and a question to help you remember what you would like to ask your instructor, your tutor, or another student.

Problem Number	Question? Where in the problem did you start to have difficulty or confusion?	Answered?

Often, you will have more questions than there is space provided here. If so, write them on notebook paper and be sure to talk to your instructor. You might ask in class or privately with the instructor.

Do You Really Know It?

Can you put into words the concepts that you learned in this section? Answer the below question from the *Writing* section in the *Study Set* in your text. Explain as if you were explaining to someone who has never taken this class before. Use notebook paper if you need more room.

Explain why the following factorization is not complete. Then finish the solution.

$$ax + ay + x + y = a(x+y) + x + y$$

You Write the Test!

If you were writing the test for this section, what would you want a student to know? Write two test questions that you think might come from this material. Write questions of various difficulty; these questions can be original or chosen from the homework. Be sure to supply the answer also!

Write these questions at the end of this chapter under the section titled *My Practice Chapter 5 Test* and the answer to each question under the section titled *My Practice Chapter 5 Test Answers*.

Reflect on the Section

Look back at the *Pre-Class Prep* section. Did the lecture explain topics that you thought were going to be challenging or confusing? _____

- Are there topics that you still have questions on from the reading or the lecture? If so, complete the following:
  I don't understand…

- Speak to your instructor in class or during office hours about these concerns.

Reflect on Your Math Attitude

This section marks the half-way point in this chapter. How are you doing? Think about this And list any concerns that you have as you continue your work in this chapter.

What steps, if needed, should you take to address your concerns?

5.6 Factoring Trinomials: Pre-Class Prep

 Are You Ready?

Complete the following problems. These review some basic skills that are needed when factoring trinomials. *All answers are found in the Pre-Class Prep Answer Section.*

1. In $x^2 - 6x - 27$, what is the coefficient of the leading term?

2. Multiply: $(x+8)(x-6)$

3. Multiply: $(2a+9)^2$

4. $x^2 - 4xy - 5y^2$ is a polynomial in how many variables?

5. Find two integers whose product is 10 and whose sum is 7.

6. Find two integers whose product is –18 and whose sum is 3.

 Reading Time!

While **reading Section 5.6**, identify the word or concept being defined. Choose from the following words (some may be used more than once or not at all). *All answers are found in the Pre-Class Prep Answer Section.*

prime trinomial	same sign	factor by grouping	key number
different signs	substitution	trial-and-check method	two binomials
Step 1	Step 3	$(A-B)^2$	$(A+B)^2$
Step 2	Step 4	Step 5	perfect-square trinomials
coefficient of the squared variable		two integers whose product is c and sum is b	

1 Factor Perfect-Square Trinomials

1. Trinomials that are squares of a binomial: _____

2. Factor $A^2 + 2AB + B^2$ as the square of a binomial: _____

3. Factor $A^2 - 2AB + B^2$ as the square of a binomial: _____

2 Factor Trinomials of the Form $x^2 + bx + c$

4. The leading coefficient of a trinomial: _____

5. Many trinomials factor into a product of this: _____

6. In factoring $x^2 + bx + c$, what to look for first: _____

7. When factoring $x^2 + bx + c$, if c is positive, the sign of the two numbers: _____

8. When factoring $x^2 + bx + c$, if c is negative, the sign of the two numbers: _____

3 Factor Trinomials of the form $ax^2 + bx + c$

9. A method used to factor trinomials of the form $ax^2 + bx + c$: _____

10. *Factoring Trinomials with Leading Coefficients Other Than 1*

   Factor out any GCF (including –1 if necessary to make a positive): _____

   Check by multiplying: _____

   Write the trinomial as a product of two binomials: _____

   Try combinations of first terms and second terms until the one that gives the correct middle

   term is found (if no combination works, the trinomial is prime): _____

   If c is positive, the signs within the binomial factors match the sign of b. If c is negative, the

   signs within the binomial factors are opposites: _____

4 Use Substitution to Factor Trinomials

11. Method to simplify complicated expressions, especially those involving a quantity within

   parentheses: _____

5 Use the Grouping Method to Factor Trinomials

12. For a trinomial of the form $ax^2 + bx + c$, the product ac: _____

13. *Factoring Trinomials by Grouping*

Express the middle term, *bx,* as the sum (or difference) of two terms, then factor the

equivalent four-term polynomial by grouping: _____

Identify *a, b,* and *c,* and find the key number *ac*: _____

Check the factorization by multiplying: _____

Find two integers whose product is the key number and whose sum is *b*: _____

Factor out any GCF (including –1 if necessary to make *a* positive): _____

✓ Getting Ready for Class

Briefly look through the section again. Answer the following by writing the concept or just the page number from the text.

Identify concepts/procedures that you feel confident about:

Identify concepts/procedures that look confusing or challenging:

Be sure to ask your instructor further questions if you are still having difficulty with a concept.

5.6 Factoring Trinomials: In-Class Notes

Terms, Definitions, and Main Ideas	Examples and Notes

Use notebook paper for additional notes

5.6 Factoring Trinomials: After Class

Important to Know

What is your homework assignment? Be sure to note it in your weekly schedule.

Section 5.6 Homework: _____ **Due**: _____

Getting to Work!

Complete your homework assignment. If you are unable to do a problem, write down the problem number and a question to help you remember what you would like to ask your instructor, your tutor, or another student.

Problem Number	Question? Where in the problem did you start to have difficulty or confusion?	Answered?

Often, you will have more questions than there is space provided here. If so, write them on notebook paper and be sure to talk to your instructor. You might ask in class or privately with the instructor.

Do You Really Know It?

Can you put into words the concepts that you learned in this section? Answer the below question from the *Writing* section in the *Study Set* in your text. Explain as if you were explaining to someone who has never taken this class before. Use notebook paper if you need more room.

How does one determine whether a trinomial is a perfect-square trinomial?

You Write the Test!

If you were writing the test for this section, what would you want a student to know? Write two test questions that you think might come from this material. Write questions of various difficulty; these questions can be original or chosen from the homework. Be sure to supply the answer also!

Write these questions at the end of this chapter under the section titled *My Practice Chapter 5 Test* and the answer to each question under the section titled *My Practice Chapter 5 Test Answers*.

Reflect on the Section

Look back at the *Pre-Class Prep* section. Did the lecture explain topics that you thought were going to be challenging or confusing? _____

- Are there topics that you still have questions on from the reading or the lecture? If so, complete the following:
    I don't understand…

- Speak to your instructor in class or during office hours about these concerns.

Reflect on Your Math Attitude

Write one word that describes how you felt when you last came to class. _____

Does your attitude reflect a strong or weak desire to learn?

What could you do to either maintain your strong desire to learn or to change your weak desire into a strong desire to learn?

Changing or maintaining your desire to learn is often an act of will. It will take commitment.

5.7 The Difference of Two Squares; the Sum and Difference of Two Cubes: Pre-Class Prep

 ## Are You Ready?

Complete the following problems. These problems review some basic skills that are needed when factoring certain types of binomials. *All answers are found in the Pre-Class Prep Answer Section.*

1. Multiply: $(n+9)(n-9)$

2. Simplify: $(8d)^2$

3. Evaluate: a. 3^3 b. 6^3

4. Simplify: $(4a)^3$

5. Multiply: $(b+4)(b^2-4b+16)$

6. Multiply: $(2a-1)(4a^2+2a+1)$

 ## Reading Time!
(on the next page)

 ## Getting Ready for Class

Briefly look through the section again. Answer the following by writing the concept or just the page number from the text.

Identify concepts/procedures that you feel confident about:

Identify concepts/procedures that look confusing or challenging:

Be sure to ask your instructor further questions if you are still having difficulty with a concept.

Reading Time!

While **reading Section 5.7**, match the word or concept to its definition or description. Not all choices are used. *All answers are found in the Pre-Class Prep Answer Section.*

1 Factor the Difference of Two Squares

_____ 1. The name for the binomial $A^2 - B^2$

_____ 2. Factored form of $A^2 - B^2$

_____ 3. Perfect-square integers

_____ 4. Expressions such as a^4 and $x^6 y^8$

_____ 5. Substitution

2 Factor the Sum and Difference of Two Cubes

_____ 6. Perfect-cube integers

_____ 7. Sum of two cubes

_____ 8. Factored form of $F^3 + L^3$

_____ 9. Difference of two cubes

_____ 10. Factored form of $F^3 - L^3$

A. perfect squares

B. $F^3 - L^3$

C. cubes of integers, such as 1, 8, 27,..., 1000

D. $(F + L)(F^2 + FL + L^2)$

E. helps simplify the factoring process

F. squares of integers, such as 1, 4, 9,...,400

G. $(A \cdot B)^2$

H. $(A - B)(A + B)$

I. $(F - L)(F^2 + FL + L^2)$

J. $F^3 + L^3$

K. $(F + L)(F^2 - FL + L^2)$

L. difference of two squares

M. $(F - L)(F^2 - FL + L^2)$

5.7 The Difference of Two Squares; the Sum and Difference of Two Cubes: In-Class Notes

Terms, Definitions, and Main Ideas	Examples and Notes

Use notebook paper for additional notes

5.7 The Difference of Two Squares; the Sum and Difference of Two Cubes: After Class

Important to Know

What is your homework assignment? Be sure to note it in your weekly schedule.

Section 5.7 Homework: _____ **Due**: _____

Getting to Work!

Complete your homework assignment. If you are unable to do a problem, write down the problem number and a question to help you remember what you would like to ask your instructor, your tutor, or another student.

Problem Number	Question? Where in the problem did you start to have difficulty or confusion?	Answered?

Often, you will have more questions than there is space provided here. If so, write them on notebook paper and be sure to talk to your instructor. You might ask in class or privately with the instructor.

Do You Really Know It?

Can you put into words the concepts that you learned in this section? Answer the below question from the *Writing* section in the *Study Set* in your text. Explain as if you were explaining to someone who has never taken this class before. Use notebook paper if you need more room.

Explain how the patterns used to factor the sum and difference of two cubes are similar and how they differ.

You Write the Test!

If you were writing the test for this section, what would you want a student to know? Write two test questions that you think might come from this material. Write questions of various difficulty; these questions can be original or chosen from the homework. Be sure to supply the answer also!

Write these questions at the end of this chapter under the section titled *My Practice Chapter 5 Test* and the answer to each question under the section titled *My Practice Chapter 5 Test Answers*.

Reflect on the Section

Look back at the *Pre-Class Prep* section. Did the lecture explain topics that you thought were going to be challenging or confusing? _____

- Are there topics that you still have questions on from the reading or the lecture? If so, complete the following:
    I don't understand…

- Speak to your instructor in class or during office hours about these concerns.

Reflect on Your Math Attitude

You have just sat down in the classroom. Are you prepared when your instructor is ready to begin class?

How would being ready with questions help to make you successful in this class?

In the *After Class* section, write your questions in the table under *Getting to Work!*.

Are You Ready?

Complete the following problems. These problems review some basic skills that are needed when factoring polynomials. *All answers are found in the Pre-Class Prep Answer Section.*

1. How many terms does each expression have?

   a. $12x^2 y^2 z^3 - 2xy^2 z^3 - 4y^2 z^3$

   b. $x^3 + 5x^2 + 6x + x^2 y + 5xy + 6y$

2. Do any of the factors in $\left(x^3 + y^3\right)\left(x^3 - y^3\right)$ factor further?

3. Multiply: $3a\left(a+5\right)\left(a-14\right)$

4. What is the greatest common factor of the terms $32m^3 n^2 - 24mn^3$?

Reading Time!

While **reading Section 5.8**, fill in the blanks choosing from the following words (some may be used more than once or not at all). *All answers are found in the Pre-Class Prep Answer Section.*

GCF	negative	check	factor(s)
difference	sum	grouping	perfect-square
terms	positive	squares	four
multiplying	cubes	trial-and-check	completely
product		polynomials	

1 Factor Random Polynomials

1. To factor a polynomial means to express it as a _____ of two (or more) _____ .

Steps for Factoring a Polynomial

2. Is there a common _____?

   If so, _____ out the _____. Factor so the leading coefficient is _____.

3. How many _____ does the polynomial have?

   If it has two terms, look for the following problem types:

   a. the _____ of two squares

   b. the _____ of two cubes

   c. the difference of two _____

   If it has three terms, look for the following problem types:

   a. a _____ trinomial

   b. if not a perfect square, use the _____ method or the _____ method.

   If it has _____ or more terms, try to factor by _____.

4. Can any _____ be factored further?

   If so, factor them _____.

5. Does the factorization _____?

   Check by _____.

✓ Getting Ready for Class

   Briefly look through the section again. Answer the following by writing the concept or just the page number from the text.

Identify concepts/procedures that you feel confident about:

Identify concepts/procedures that look confusing or challenging:

Be sure to ask your instructor further questions if you are still having difficulty with a concept.

Terms, Definitions, and Main Ideas

Examples and Notes

Use notebook paper for additional notes

5.8 Summary of Factoring Techniques: After Class

Important to Know

What is your homework assignment? Be sure to note it in your weekly schedule.

Section 5.8 Homework: _____ **Due**: _____

Getting to Work!

Complete your homework assignment. If you are unable to do a problem, write down the problem number and a question to help you remember what you would like to ask your instructor, your tutor, or another student.

Problem Number	Question? Where in the problem did you start to have difficulty or confusion?	Answered?

Often, you will have more questions than there is space provided here. If so, write them on notebook paper and be sure to talk to your instructor. You might ask in class or privately with the instructor.

Do You Really Know It?

Can you put into words the concepts that you learned in this section? Answer the below question from the *Writing* section in the *Study Set* in your text. Explain as if you were explaining to someone who has never taken this class before. Use notebook paper if you need more room.

What is your strategy for factoring a polynomial?

You Write the Test!

If you were writing the test for this section, what would you want a student to know? Write two test questions that you think might come from this material. Write questions of various difficulty; these questions can be original or chosen from the homework. Be sure to supply the answer also!

Write these questions at the end of this chapter under the section titled *My Practice Chapter 5 Test* and the answer to each question under the section titled *My Practice Chapter 5 Test Answers*.

Reflect on the Section

Look back at the *Pre-Class Prep* section. Did the lecture explain topics that you thought were going to be challenging or confusing? _____

- Are there topics that you still have questions on from the reading or the lecture? If so, complete the following:
    I don't understand…

- Speak to your instructor in class or during office hours about these concerns.

Reflect on Your Math Attitude

Do you usually like to strategize, or would you rather just jump in with no set plan?

In this section, a specific strategy for factoring was presented. How do you feel about using this strategy? Does having a strategy raise your comfort level with factoring? Why or why not?

Writing out the factoring strategy will not only be a good reference, but it will also reinforce the steps to successful factoring.

5.9 Solving Equations by Factoring: Pre-Class Prep

 Are You Ready?

Complete the following problems. These problems review some basic skills that are needed when solving quadratic equations. *All answers are found in the Pre-Class Prep Answer Section.*

1. Fill in the blank: $8 \cdot \underline{} = 0$

2. Solve: $x + 4 = 0$

3. Solve: $8x = 0$

4. Factor: $x^2 - x - 6$

5. Factor: $3n^2 - n - 2$

6. Factor: $x^3 - x^2 - 121x + 121$

 Reading Time!

While **reading Section 5.9**, identify the word or concept being defined. Choose from the following words/concepts (some may be used more than once or not at all). *All answers are found in the Pre-Class Prep Answer Section.*

Check the results in the original equation.	Factor the polynomial.
$ax^2 + bx + c = 0$, a, b, c are real numbers, $a \neq 0$	three solutions
extension of the zero-factor property	Zero-Factor Property
Write the equation in standard form.	Solve each resulting equation.
two solutions	should be discarded or rejected
solution of a quadratic equation	polynomial equation
Use the zero-factor property to set each factor equal to zero.	

1 Solve Quadratic Equations Using the Zero-Factor Property

1. Two polynomials set equal to each other: _____

2. A quadratic equation in standard form: _____

3. Value of the variable in a quadratic equation that makes the equation true: _____

4. If the product of two real numbers is 0, at least one of them is 0: _____

5. First step of the factoring method for solving a quadratic equation: _____

6. Second step of the factoring method for solving a quadratic equation: _____

7. Third step of the factoring method for solving a quadratic equation: _____

8. Fourth step of the factoring method for solving a quadratic equation: _____

9. Last step of the factoring method for solving a quadratic equation: _____

2 Solve Higher-Degree Polynomial Equations by Factoring

10. When the product of two or more real numbers is 0, at least one of them is 0: _____

11. The number of solutions for the third-degree equation $6x^3 - x^2 = 2x$: _____

3 Use Quadratic Equations to Solve Problems

12. A solution of a quadratic equation that does not make sense in the real-world: _____

✓ Getting Ready for Class

Briefly look through the section again. Answer the following by writing the concept or just the page number from the text.

Identify concepts/procedures that you feel confident about:

Identify concepts/procedures that look confusing or challenging:

Be sure to ask your instructor further questions if you are still having difficulty with a concept.

5.9 Solving Equations by Factoring: In-Class Notes

Terms, Definitions, and Main Ideas	Examples and Notes

Use notebook paper for additional notes

5.9 Solving Equations by Factoring: After Class

Important to Know

What is your homework assignment? Be sure to note it in your weekly schedule.

Section 5.9 Homework: _____ **Due**: _____

Getting to Work!

Complete your homework assignment. If you are unable to do a problem, write down the problem number and a question to help you remember what you would like to ask your instructor, your tutor, or another student.

Problem Number	Question? Where in the problem did you start to have difficulty or confusion?	Answered?

Often, you will have more questions than there is space provided here. If so, write them on notebook paper and be sure to talk to your instructor. You might ask in class or privately with the instructor.

Do You Really Know It?

Can you put into words the concepts that you learned in this section? Answer the below question from the *Writing* section in the *Study Set* in your text. Explain as if you were explaining to someone who has never taken this class before. Use notebook paper if you need more room.

Explain what is wrong with the following solution.
  Solve:

You Write the Test!

If you were writing the test for this section, what would you want a student to know? Write two test questions that you think might come from this material. Write questions of various difficulty; these questions can be original or chosen from the homework. Be sure to supply the answer also!

Write these questions at the end of this chapter under the section titled *My Practice Chapter 5 Test* and the answer to each question under the section titled *My Practice Chapter 5 Test Answers*.

Reflect on the Section

Look back at the *Pre-Class Prep* section. Did the lecture explain topics that you thought were going to be challenging or confusing? _____

- Are there topics that you still have questions on from the reading or the lecture? If so, complete the following:
    I don't understand…

- Speak to your instructor in class or during office hours about these concerns.

Reflect on Your Math Attitude

How comfortable are you with the material presented in this section?

What could you do to either improve your comfort level or to make sure that you continue to stay comfortable with the material?

Be sure to utilize the online resources to help you with the concepts presented in this section.

Chapter 5 Activities

 Your instructor may assign these activities to you to complete in class, or you may complete them on your own to solidify your understanding of chapter topics. The activities begin on the next page.

❖ **Student Activity:** *Visual Representation for Multiplying Binomials*
   Represent the multiplication of two binomials through a diagram called an area model.

❖ **Student Activity:** *Skeletons of Tricky Binomials*
   Categorize a polynomial expression; write the appropriate skeleton, and then factor.

❖ **Guided Learning Activity:** *Vertical Form of Polynomial Multiplication*
   Multiply polynomials 'vertically.'

Student Activity
Visual Representation for Multiplying Binomials

Multiplication can be represented using a diagram called an area model. Find an expression for the area of each of the squares and rectangles below. Remember that the formula for finding the area of a rectangle is Area = (length)(width). Write the expression for the area inside each square or rectangle, and then write it inside the same square or rectangle in the composite figure below. **Thus each area will be written in two places.**

Problem 1:

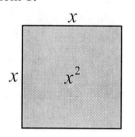 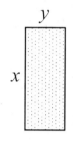 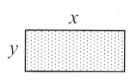

To find the area of the composite rectangle, we could either add the areas of all the smaller pieces, which would give this area:

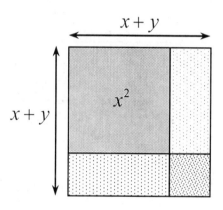

_____

Or we could find the area by using the area formula, which would give $\text{Area} = (x+y)(x+y) = (x+y)^2$

This tells us that $(x+y)^2 = $ _____ .

Problem 2: Do this one the same way.

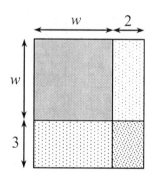 

What mathematical equation does this set of figures tell us?

_____ = _____

Cengage Student Workbook Activities, M. Andersen

Student Activity
Skeletons of Tricky Binomials

Directions: Categorize each of the expressions using the choices below, then write the appropriate skeleton for the problem, and factor the expression. If the expression cannot be factored, say so. The first one has been done for you.

A Sum of Squares: $(\ )^2 + (\ )^2$

B Difference of Squares: $(\ )^2 - (\ )^2$

C Sum of Cubes: $(\ )^3 + (\ )^3$

D Difference of Cubes: $(\ )^3 - (\ )^3$

E None of these

Polynomial	Category	Skeleton	Factored Form
1. $x^4 - (a+b)^2$	B	$(x^2)^2 - (a+b)^2$	$(x^2 + (a+b))(x^2 - (a+b))$ or $(x^2 + a + b)(x^2 - a - b)$
1. $(a-b)^2 - x^4$			
2. $1 + (x+y)^3$			
3. $(a-b)^2 + x^2$			
4. $\dfrac{8}{27} - x^6$			
5. $x^3 - 36$			
6. $x^2 + 2xy + y^2 - 100$			
7. $(x+y)^2 + 27$			
8. $(a+b)^3 - (c-d)^3$			

Cengage Student Workbook Activities, M. Andersen

Guided Learning Activity
Vertical Form of Polynomial Multiplication

The expressions $304 \cdot 21$ and $(3x^2 + 4) + (2x + 1)$ are simplified in almost the same way: For both expressions, we line up the like terms vertically; then multiply.

Example 1:

```
      3  0  4              3 hundreds  0 tens  4 ones                  3x²  +0x  +4
 ×       2  1         ×                2 tens  1 one         ×              +2x  +1
 ─────────────        ───────────────────────────────       ──────────────────────
      3  0  4              3 hundreds  0 tens  4 ones                 +3x²  +0x  +4
   6  0  8            6 thousands  0 hundreds  8 tens              6x³  +0x²  +8x
 ─────────────        ───────────────────────────────       ──────────────────────
   6  3  8  4         6 thousands  3 hundreds  8 tens  4 ones      6x³  +3x²  +8x  +4
```

One of the differences between the vertical multiplications of the expressions is that we cannot "carry" coefficients greater than 9 to the next column like we carry numbers greater than 9. Another difference is that we can have negative coefficients if that is part of the original expression.

Example 2: $(4x + 5)(2x - 6)$ becomes

```
              4x     +5
     ×        2x     -6
     ──────────────────
             -24x    -30
     8x²     +10x
     ──────────────────
     8x²     -14x    -30
```

Now try these. Make sure to **line up the like terms** in the same columns and keep the signed coefficients with the terms in the columns.

a. $(7y - 6)(y + 5)$

b. $(x^2 - 3x + 6)(x - 9)$

c. $(a^2 + 8a + 6)(a^2 + 5)$

d. $(2x^2 + 5x - 4)(x^2 - 6x + 7)$

Chapter 5 Test Skills Assessment

Pre-Test Preparation Work:

1. Re-read the objectives from each section.
2. Review the *Reading Time!* activity for each section.
3. Go over all your classroom notes, if something in your notes doesn't make sense to you, make a note and ask your teacher or a classmate.
4. Make additional notations to your work if your teacher states specific concepts to study in preparation for the chapter test.
5. Attend any study sessions held by your teacher or teaching assistant.
6. Practice additional problems.
7. Go over any missed problems in your homework sets.
8. Talk out concepts with your peers in small group study sessions.

List other preparations that you have found beneficial in preparing for a math test.

Additional Practice Suggestions

1. Use your book's review problems at the back of the chapter as a practice test.
2. Take *My Practice Chapter Test* and the text's *Chapter Test*. Time yourself and do not use your notebook or textbook.
3. Pace yourself as you work through these problems.
4. Read each question carefully, playing close attention to the instructions.
5. Check your work using the answers provided.
6. Rework any missed problems. Do not just "look them over" but actually rework the problem without looking at text or notes.

My Practice Chapter 5 Test

For each section, you had the opportunity to create two test questions under the section *You Write the Test!* Write each of those questions here. Include your answers under the heading *My Practice Chapter 5 Test Answers*. **Take the test without notes or your textbook.** If you do not get a question correct, review the text and/or your notes then take the test again. For further review, do the *Chapter 5 Test* in the text.

Section 5.1

1.

2.

Section 5.2

3.

4.

Section 5.3

5.

6.

Section 5.4

7.

8.

Section 5.5

9.

10.

Section 5.6

11.

12.

Section 5.7

13.

14.

Section 5.8

15.

16.

Section 5.9

17.

18.

My Practice Chapter 5 Test Answers

Section 5.1

1.

2.

Section 5.2

3.

4.

Section 5.3

5.

6.

Section 5.4

7.

8.

Section 5.5

9.

10.

Section 5.6

11.

12.

Section 5.7

13.

14.

Section 5.8

15.

16.

Section 5.9

17.

18.

Chapter 6 Rational Expressions and Equations

Read the *Study Skills Workshop* found at the beginning of Chapter 6 in your textbook. **Complete** the activities below for this chapter's *Study Skills Workshop*.

Reading the Textbook

One is not able to read an algebra textbook in the same way as a newspaper or a novel. Often understanding is not immediate and must be obtained through careful reading, a step-by-step process.

Do you read your algebra textbook? YES   NO   SOMETIMES

✓ Identify some difficulties you (or another student) might have had reading your algebra textbook.

_____

_____

Below are some strategies to use when reading your textbook.

Skimming for an Overview

To become familiar with new vocabulary and notation that will be used by your instructor, take a quick look at the material before class. For each section, this notebook has provided a *Reading Time!* section to help you identify new vocabulary, concepts, and notation.

✓ For each section, complete the ***Reading Time!*** section in this notebook.

Reading for Understanding

To read for understanding, one needs to not only read slowly but with a pencil and paper at hand. Do not skip anything when you read as every word is important!

✓ Read each section slowly. Highlight important points.

✓ Work each *Self Check* problem. Your solutions should look like those in the *Examples*.

✓ In the ***Getting Ready for Class*** section in this notebook,
   o Identify concepts/procedures that you feel confident about
   o Identify concepts/procedures that look confusing or challenging

✓ Be alert during class when the instructor addresses the concepts.

6.1 Rational Functions and Simplifying Rational Expressions: Pre-Class Prep

 Are You Ready?

Complete the following problems. These review some basic skills when simplifying rational expressions. *All answers are found in the Pre-Class Prep Answer Section.*

1. Evaluate: a. $\dfrac{0}{10}$ b. $\dfrac{10}{0}$ c. $\dfrac{9}{-9}$ 2. Simplify: $\dfrac{36}{28}$

3. Factor: $12a^5 - 16a^4$ 4. Factor: $x^2 - 100$

5. Factor: $5n^2 + 23n - 10$ 6. Factor: $2x^2 + x^2 - 14x - 7$

 Reading Time!

While **reading Section 6.1**, identify the word or concept being defined. Choose from the following words (some may be used more than once or not at all). *All answers are found in the Pre-Class Prep Answer Section.*

rational function	rational expression	vertical asymptote
evaluate an expression	first step to simplify rational expressions	-1
reciprocal function	simplify a rational expression	horizontal asymptote
opposites	second step to simplify rational expressions	domain
1	equivalent expressions	asymptote
$c(2) = \dfrac{1.50(2)+6}{2} = 4.5$	$\dfrac{a}{b} = \dfrac{c}{d}$ if and only if $ad = bc$	$\dfrac{ak}{bk} = \dfrac{a}{b} \cdot \dfrac{k}{k} = \dfrac{a}{b}$
$\dfrac{a}{1} = a$ and $\dfrac{a}{a} = 1$	$-\dfrac{a}{b} = \dfrac{-a}{b} = \dfrac{a}{-b}$	$2c = \dfrac{1.50(2)+6}{2} = 4.5$

1 Define Rational Expressions and Rational Functions

1. An expression of the form $\dfrac{A}{B}$, where A and B are polynomials and B does not equal 0:

   _____

2. A function whose equation is defined by a rational expression in one variable, where the value of the polynomial in the denominator is never zero: _____

2 Evaluate Rational Functions

3. Given rational function $c(n) = \dfrac{1.50n + 6}{n}$, determine $c(2)$: _____

3 Find the Domain of a Rational Function

4. Set of all permissible input values for the variable in a function: _____

3 Recognize the Graphs of Rational Functions

5. A name for the simplest of rational functions, $f(x) = \dfrac{1}{x}$: _____

6. A line that a graph approaches: _____

7. For the graph, $f(x) = \dfrac{1}{x}$, the y-axis:_____

8. For the graph, $f(x) = \dfrac{1}{x}$, the x-axis:_____

5 Simplify Rational Expressions

9. Four properties of fractions: _____

10. Factor the numerator and denominator completely to determine their common factors:

    _____

11. To write the numerator and denominator of a rational expression such that they have no common factors other than 1:_____

12. The original rational expression and the simplified expression, which have the same value:

    _____

Chapter 6 Rational Expressions and Equations

6 Simplify Rational Expressions that have Factors that are Opposites

13. The quotient of any nonzero polynomial and its opposite, such as $\dfrac{a-b}{b-a} = ?$: _____

14. Polynomials with terms that are the same, yet opposite in sign:_____

✓ **Getting Ready for Class**

    Briefly look through the section again. Answer the following by writing the concept or just the page number from the text.

Identify concepts/procedures that you feel confident about:

Identify concepts/procedures that look confusing or challenging:

Be sure to ask your instructor further questions if you are still having difficulty with a concept.

6.1 Rational Functions and Simplifying Rational Expressions: In-Class Notes

Terms, Definitions, and Main Ideas	Examples and Notes

Use notebook paper for additional notes

6.1 Rational Functions and Simplifying Rational Expressions: After Class

 Important to Know

What is your homework assignment? Be sure to note it in your weekly schedule.

Section 6.1 Homework: _____ **Due:** _____

 Getting to Work!

Complete your homework assignment. If you are unable to do a problem, write down the problem number and a question to help you remember what you would like to ask your instructor, your tutor, or another student.

Problem Number	Question? Where in the problem did you start to have difficulty or confusion?	Answered?

Often, you will have more questions than there is space provided here. If so, write them on notebook paper and be sure to talk to your instructor. You might ask in class or privately with the instructor.

Do You Really Know It?

Can you put into words the concepts that you learned in this section? Answer the below question from the *Writing* section in the *Study Set* in your text. Explain as if you were explaining to someone who has never taken this class before. Use notebook paper if you need more room.

What does it mean when we say that $\dfrac{3x-12}{3x+15}$ and $\dfrac{x-4}{x+5}$ are equivalent expressions?

You Write the Test!

If you were writing the test for this section, what would you want a student to know? Write two test questions that you think might come from this material. Write questions of various difficulty; these questions can be original or chosen from the homework. Be sure to supply the answer also!

Write these questions at the end of this chapter under the section titled *My Practice Chapter 6 Test* and the answer to each question under the section titled *My Practice Chapter 6 Test Answers*.

Reflect on the Section

Look back at the *Pre-Class Prep* section. Did the lecture explain topics that you thought were going to be challenging or confusing? _____

- Are there topics that you still have questions on from the reading or the lecture? If so, complete the following:
  I don't understand...

- Speak to your instructor in class or during office hours about these concerns.

Reflect on Your Math Attitude

How often do you read your math textbook? Circle one.

Always   Sometimes   Never

How would reading your textbook regularly benefit you? How might it affect your confidence as you move into new material?

Use the *Reading Time!* section in this notebook to help you develop a habit of reading your math text.

6.2 Multiplying and Dividing Rational Expressions: Pre-Class Prep

Are You Ready?

Complete the following problems. These review some basic skills that are needed when multiplying and dividing rational expressions. *All answers are found in the Pre-Class Prep Answer Section.*

1. Multiply: $\dfrac{3}{8} \cdot \dfrac{1}{7}$

2. What is the reciprocal of $\dfrac{2}{3}$?

3. Divide: $\dfrac{21}{25} \div \dfrac{7}{15}$

4. Factor: $x^3 - x^4$

5. Simplify: $\dfrac{x(x+5)(x-7)}{4x(x-5)(x-7)}$

6. Simplify: $\dfrac{x^3 - 125}{2x - 10}$

Reading Time!

(on the next page)

Getting Ready for Class

Briefly look through the section again. Answer the following by writing the concept or just the page number from the text.

Identify concepts/procedures that you feel confident about:

Identify concepts/procedures that look confusing or challenging:

Be sure to ask your instructor further questions if you are still having difficulty with a concept.

Reading Time!

While **reading Section 6.2**, match the word or concept to its definition or description. Not all choices are used. *All answers are found in the Pre-Class Prep Answer Section.*

1 Multiply Rational Expressions

_____ 1. Multiplying Rational Expressions

_____ 2. To prepare to simplify the product of rational expressions

2 Find Powers of Rational Expressions

_____ 3. To simplify $\left(\dfrac{x^2+x-1}{2x+3}\right)^2$, rewrite as...

3 Divide Rational Expressions

_____ 4. Product of a number and its reciprocal

_____ 5. Dividing Rational Expressions

_____ 6. $\dfrac{b^3-4b}{x-1} \div (b-2)$, rewritten as a multiplication problem

4 Perform Mixed Operations

_____ 7. Order of operation rule

_____ 8. Fraction to be inverted in the problem:

$$\frac{x^2+2x-3}{6x^2+5x+1} \div \frac{2x^2-2}{2x^2-5x-3} \cdot \frac{6x^2+4x-2}{x^2-2x-3}$$

A. $\dfrac{2x^2-2}{2x^2-5x-3}$

B. invert the fraction

C. $\dfrac{b^3-4b}{x-1} \cdot \dfrac{1}{(b-2)}$

D. $\dfrac{b^3-4b}{x-1} \cdot \dfrac{(b-2)}{1}$

E. $\dfrac{6x^2+4x-2}{x^2-2x-3}$

F. factor the numerator and factor the denominator

G. $\dfrac{x^2+2x-3}{6x^2+5x+1}$

H. For any two rational expressions, $\dfrac{A}{B}$ and $\dfrac{C}{D}$, where $\dfrac{C}{D} \neq 0$, $\dfrac{A}{B} \div \dfrac{C}{D} = \dfrac{A}{B} \cdot \dfrac{D}{C} = \dfrac{AD}{BC}$.

I. perform division and multiplication in order from left to right

J. For any two rational expressions, $\dfrac{A}{B}$ and $\dfrac{C}{D}$, $\dfrac{A}{B} \cdot \dfrac{C}{D} = \dfrac{AC}{BD}$.

K. $\left(\dfrac{x^2+x-1}{2x+3}\right) \cdot \left(\dfrac{x^2+x-1}{2x+3}\right)$

L. 1

Chapter 6 Rational Expressions and Equations

6.2 Multiplying and Dividing Rational Expressions: In-Class Notes

Terms, Definitions, and Main Ideas	Examples and Notes

Use notebook paper for additional notes

6.2 Multiplying and Dividing Rational Expressions: After Class

Important to Know

What is your homework assignment? Be sure to note it in your weekly schedule.

Section 6.2 Homework: _____ **Due**: _____

Getting to Work!

Complete your homework assignment. If you are unable to do a problem, write down the problem number and a question to help you remember what you would like to ask your instructor, your tutor, or another student.

Problem Number	Question? Where in the problem did you start to have difficulty or confusion?	Answered?

Often, you will have more questions than there is space provided here. If so, write them on notebook paper and be sure to talk to your instructor. You might ask in class or privately with the instructor.

Do You Really Know It?

Can you put into words the concepts that you learned in this section? Answer the below question from the *Writing* section in the *Study Set* in your text. Explain as if you were explaining to someone who has never taken this class before. Use notebook paper if you need more room.

Explain how to multiply two rational expressions.

You Write the Test!

If you were writing the test for this section, what would you want a student to know? Write two test questions that you think might come from this material. Write questions of various difficulty; these questions can be original or chosen from the homework. Be sure to supply the answer also!

Write these questions at the end of this chapter under the section titled *My Practice Chapter 6 Test* and the answer to each question under the section titled *My Practice Chapter 6 Test Answers*.

Reflect on the Section

Look back at the *Pre-Class Prep* section. Did the lecture explain topics that you thought were going to be challenging or confusing? _____

- Are there topics that you still have questions on from the reading or the lecture? If so, complete the following:
    I don't understand...

- Speak to your instructor in class or during office hours about these concerns.

Reflect on Your Math Attitude

How comfortable are you with this section?

How would reading your math textbook make you feel even more comfortable?

The textbook provides additional examples of the concepts to help you master the material.

6.3 Adding and Subtracting Rational Expressions: Pre-Class Prep

 Are You Ready?

Complete the following problems. These review some basic skills that are needed when adding and subtracting rational expressions. *All answers are found in the Pre-Class Prep Answer Section.*

1. Add: $\dfrac{5}{11}+\dfrac{3}{11}$

2. Subtract: $\dfrac{2}{3}-\dfrac{1}{5}$

3. Simplify: $x^2+9x-(6x-4)$

4. Simplify: $\dfrac{x+9}{x+9}$

5. Simplify: $\dfrac{2a+14}{(a+7)(a-2)}$

6. Factor: $5x^2-5$

 Reading Time!

While **reading Section 6.3**, identify the statements as True or False. *All answers are found in the Pre-Class Prep Answer Section.*

1 Add and Subtract Rational Expressions with Like Denominators

_____ 1. To add rational expressions that have the same denominator, add their numerators and write the sum over the common denominator.

_____ 2. To subtract rational expressions, subtract the numerators and subtract the denominators.

2 Find the Least Common Denominator

_____ 3. The first step in finding the LCD is to factor each denominator completely.

_____ 4. The LCD is a sum that uses each different factor of the denominators the greatest number of times it appears in any one factorization.

Chapter 6 Rational Expressions and Equations

3 Add and Subtract Rational Expressions with Unlike Denominators

_____ 5. To build a rational expression, multiply it by 1 in the form of $\dfrac{c}{c}$, where c is any nonzero number or expression.

_____ 6. To add or subtract rational expressions with unlike denominators, first find the LCD.

_____ 7. All results of the addition or subtraction of rational expressions can be further simplified.

4 Add and Subtract Rational Expressions that have Denominators that are Opposites

_____ 8. When a polynomial is multiplied by $\dfrac{-1}{-1}$, the result is its opposite.

_____ 9. To add or subtract rational expressions whose denominators are opposites, multiply either expression by 1 in the form of $\dfrac{-1}{-1}$ to obtain a common denominator.

5 Perform Mixed Operations

_____ 10. To perform a combination of additions and subtractions, first find the LCD of all rational expressions.

Getting Ready for Class

Briefly look through the section again. Answer the following by writing the concept or just the page number from the text.

Identify concepts/procedures that you feel confident about:

Identify concepts/procedures that look confusing or challenging:

Be sure to ask your instructor further questions if you are still having difficulty with a concept.

6.3 Adding and Subtracting Rational Expressions: In-Class Notes

Terms, Definitions, and Main Ideas	Examples and Notes

Use notebook paper for additional notes.

6.3 Adding and Subtracting Rational Expressions: After Class

Important to Know

What is your homework assignment? Be sure to note it in your weekly schedule.

Section 6.3 Homework: _____ **Due**: _____

Getting to Work!

Complete your homework assignment. If you are unable to do a problem, write down the problem number and a question to help you remember what you would like to ask your instructor, your tutor, or another student.

Problem Number	Question? Where in the problem did you start to have difficulty or confusion?	Answered?

Often, you will have more questions than there is space provided here. If so, write them on notebook paper and be sure to talk to your instructor. You might ask in class or privately with the instructor.

Do You Really Know It?

Can you put into words the concepts that you learned in this section? Answer the below question from the *Writing* section in the *Study Set* in your text. Explain as if you were explaining to someone who has never taken this class before. Use notebook paper if you need more room.

Explain how to find the least common denominator of a set of rational expressions.

You Write the Test!

If you were writing the test for this section, what would you want a student to know? Write two test questions that you think might come from this material. Write questions of various difficulty; these questions can be original or chosen from the homework. Be sure to supply the answer also!

Write these questions at the end of this chapter under the section titled *My Practice Chapter 6 Test* and the answer to each question under the section titled *My Practice Chapter 6 Test Answers*.

Reflect on the Section

Look back at the *Pre-Class Prep* section. Did the lecture explain topics that you thought were going to be challenging or confusing? _____

- Are there topics that you still have questions on from the reading or the lecture? If so, complete the following:
  I don't understand...

- Speak to your instructor in class or during office hours about these concerns.

Reflect on Your Math Attitude

Adding and subtracting fractions with unlike denominators is often a challenge for students. Has it been a challenge for you?

Now, you are working with rational expressions. Have you seen the similarities to numerical fractions? How has that made you feel?

Even though the concepts are different, use prior knowledge to make them more familiar.

6.4 Simplifying Complex Fractions: Pre-Class Prep

 Are You Ready?

Complete the following problems. These review some basic skills that are needed when simplifying complex fractions. *All answers are found in the Pre-Class Prep Answer Section.*

1. What operation is indicated by the fraction bar in $\dfrac{64}{16}$?   2. Divide: $\dfrac{2}{21} \div \dfrac{8}{27}$

3. What is the LCD of $\dfrac{5}{a-2}$ and $\dfrac{3}{a+1}$?   4. Multiply: $\left(\dfrac{3a}{5} - \dfrac{1}{a}\right)20a$

 Reading Time!

While **reading Section 6.4**, fill in the blanks choosing from the following words (some may be used more than once or not at all). *All answers are found in the Pre-Class Prep Answer Section.*

complex fraction	single	LCD	division	complex
numerator	1	x	all	reciprocal
multiplication	denominator	$x+3$	no	simplify
$\dfrac{3a}{b} \cdot \dfrac{b^2}{6ac}$		$\dfrac{3a}{b} \cdot \dfrac{6ac}{b^2}$		

1 Simplify Complex Fractions Using Division

1. A rational expression whose numerator and/or denominator contain rational expressions is

   called a _____ rational expression, or a _____.

2. Simplifying Complex Fractions Method 1: Using _____

   Step 1: Add or subtract in the _____ and/or denominator so that the numerator is a

   _____ rational expression and the denominator is a _____ rational

   expression.

Step 2: Perform the indicated _____ by multiplying the numerator of the complex fraction by the _____ of the denominator.

Step 3: _____ the result, if possible.

3. The complex fraction, $\dfrac{\dfrac{3a}{b}}{\dfrac{6ac}{b^2}}$, written as a multiplication problem is _____.

2 Simplify Complex Fractions Using the LCD

4. Simplifying Complex Fractions Method 2: Multiplying by the _____

Step 1: Find the _____ of _____ rational expressions within the complex fraction.

Step 2: Multiply the complex fraction by _____ in the form $\dfrac{\text{LCD}}{\text{LCD}}$.

Step 3: Perform the operations in the numerator and denominator. _____ rational expressions should remain within the complex fraction.

Step 4: _____ the result, if possible.

5. The LCD of the complex fraction, $\dfrac{\dfrac{2}{x}+5}{x+3}$, is _____.

Getting Ready for Class

Briefly look through the section again. Answer the following by writing the concept or just the page number from the text.

Identify concepts/procedures that you feel confident about:

Identify concepts/procedures that look confusing or challenging:

Be sure to ask your instructor further questions if you are still having difficulty with a concept.

Chapter 6 Rational Expressions and Equations

6.4 Simplifying Complex Fractions: In-Class Notes

Terms, Definitions, and Main Ideas	Examples and Notes

Use notebook paper for additional notes

6.4 Simplifying Complex Fractions: After Class

Important to Know

What is your homework assignment? Be sure to note it in your weekly schedule.

Section 6.4 Homework: _____ Due: _____

Getting to Work!

Complete your homework assignment. If you are unable to do a problem, write down the problem number and a question to help you remember what you would like to ask your instructor, your tutor, or another student.

Problem Number	Question? Where in the problem did you start to have difficulty or confusion?	Answered?

Often, you will have more questions than there is space provided here. If so, write them on notebook paper and be sure to talk to your instructor. You might ask in class or privately with the instructor.

Do You Really Know It?

Can you put into words the concepts that you learned in this section? Answer the below question from the *Writing* section in the *Study Set* in your text. Explain as if you were explaining to someone who has never taken this class before. Use notebook paper if you need more room.

What does it mean when we say that $\dfrac{\dfrac{1}{x+2}}{1+\dfrac{1}{x+2}}$ *and* $\dfrac{1}{x+3}$ *are equivalent expressions?*

Chapter 6 Rational Expressions and Equations

You Write the Test!

If you were writing the test for this section, what would you want a student to know? Write two test questions that you think might come from this material. Write questions of various difficulty; these questions can be original or chosen from the homework. Be sure to supply the answer also!

Write these questions at the end of this chapter under the section titled *My Practice Chapter 6 Test* and the answer to each question under the section titled *My Practice Chapter 6 Test Answers*.

Reflect on the Section

Look back at the *Pre-Class Prep* section. Did the lecture explain topics that you thought were going to be challenging or confusing? _____

- Are there topics that you still have questions on from the reading or the lecture? If so, complete the following:
    I don't understand...

- Speak to your instructor in class or during office hours about these concerns.

Reflect on Your Math Attitude

In this section you were presented with two different methods of simplifying a complex fraction. Describe how that makes you feel when you are given two different ways to do something.

Do you feel that it will be worth your effort to learn both ways, or do you want to just stay with one method? Why?

In math, there are often different paths to get to the same answer.

6.5 Dividing Polynomials: Pre-Class Prep

 Are You Ready?

Complete the following problems. These problems review some basic skills that are needed when dividing polynomials. *All answers are found in the Pre-Class Prep Answer Section.*

1. Simplify: a. $\dfrac{12}{30}$

   b. $\dfrac{8a^8}{4a^3}$

2. Multiply: $4x^2y^3\left(2x^3y^2 - 5x^2y^4 + \dfrac{1}{2xy}\right)$

3. Divide: $24\overline{)864}$

4. Subtract:
$$\begin{array}{r} 4x^3 + 7x^2 \\ -\left(4x^3 + 5x^2\right) \\ \hline \end{array}$$

 Reading Time!

While **reading Section 6.5**, fill in the blanks choosing from the following words (some may be used more than once or not at all). *All answers are found in the Pre-Class Prep Answer Section.*

1	positive	quotient rule	simplifying fractions	missing
0	simplest	long division	dividend	descending
divide	placeholder	less	quotient	monomial
$\dfrac{A}{D} + \dfrac{B}{D}$	remainder	subtract	common	$\dfrac{A}{D} + B$
ascending	divisor	add	term	multiply

1 Divide a Monomial by a Monomial

1. To _____ monomials, use the method for _____ or the _____ for exponents.

2. When dividing a monomial by a monomial, write answers using _____ exponents.

2 Divide a Polynomial by a Monomial

3. To divide a polynomial by a _____, divide each _____ of the polynomial by the monomial.

4. $\dfrac{A+B}{D} =$ _____

5. Only factors_____ to the numerator and denominator can be removed.

3 Divide a Polynomial by a Polynomial

6. Use a method similar to _____ in arithmetic to divide a polynomial by a polynomial (other than a monomial).

7. For any division: _____ · _____ + _____ = _____

8. The _____ method is a series of four steps: _____, multiply, _____, bring down the next term.

9. *Dividing a Polynomial by a Polynomial*

   Step 1: To begin, arrange the terms of the polynomials in _____ powers of the variable.

   Step 2: Perform the long division process until the degree of the remainder is _____ than the degree of the _____.

   If there is a remainder, write the result in the form: _____ + $\dfrac{\rule{1cm}{0.4pt}}{divisor}$.

4 Divide Polynomials with Missing Terms

10. When a power of the variable is _____ in the dividend, insert _____ terms with a coefficient of _____.

Getting Ready for Class

Briefly look through the section again. Answer the following by writing the concept or just the page number from the text.

Identify concepts/procedures that you feel confident about:

Identify concepts/procedures that look confusing or challenging:

Be sure to ask your instructor further questions if you are still having difficulty with a concept.

Chapter 6 Rational Expressions and Equations

6.5 Dividing Polynomials: In-Class Notes

Terms, Definitions, and Main Ideas	Examples and Notes

Use notebook paper for additional notes

6.5 Dividing Polynomials: After Class

Important to Know

What is your homework assignment? Be sure to note it in your weekly schedule.

Section 6.5 Homework: _____ **Due**: _____

Getting to Work!

Complete your homework assignment. If you are unable to do a problem, write down the problem number and a question to help you remember what you would like to ask your instructor, your tutor, or another student.

Problem Number	Question? Where in the problem did you start to have difficulty or confusion?	Answered?

Often, you will have more questions than there is space provided here. If so, write them on notebook paper and be sure to talk to your instructor. You might ask in class or privately with the instructor.

Do You Really Know It?

Can you put into words the concepts that you learned in this section? Answer the below question from the *Writing* section in the *Study Set* in your text. Explain as if you were explaining to someone who has never taken this class before. Use notebook paper if you need more room.

Explain how to check to determine if $\left(3x^2 - 15\right) \div \left(x + 3\right) = 3x - 9 + \dfrac{12}{x+3}$.

You Write the Test!

If you were writing the test for this section, what would you want a student to know? Write two test questions that you think might come from this material. Write questions of various difficulty; these questions can be original or chosen from the homework. Be sure to supply the answer also!

Write these questions at the end of this chapter under the section titled *My Practice Chapter 6 Test* and the answer to each question under the section titled *My Practice Chapter 6 Test Answers*.

Reflect on the Section

Look back at the *Pre-Class Prep* section. Did the lecture explain topics that you thought were going to be challenging or confusing? _____

- Are there topics that you still have questions on from the reading or the lecture? If so, complete the following:
  I don't understand...

- Speak to your instructor in class or during office hours about these concerns.

Reflect on Your Math Attitude

How did that make you feel when you learned that dividing polynomials was similar to long division done in arithmetic?

Think of other instances when what seemed new was actually something old, just in a different form? How does that knowledge affect your attitude toward the new topic?

6.6 Synthetic Division: Pre-Class Prep

 Are You Ready?

Complete the following problems. These review some basic skills that are needed when using synthetic division. *All answers are found in the Pre-Class Prep Answer Section.*

1. Divide: $x+5\overline{)3x^2+19x+20}$

2. Divide: $\dfrac{2x^2+x-9}{x-2}$

3. Let $f(x)=x^4+3x^3-x-12$. Find $f(x)$.

4. Subtract:
$$9x^3 + 0x^2$$
$$-\left(9x^3-27x^2\right)$$

 Reading Time!

(on the next page)

 Getting Ready for Class

Briefly look through the section again. Answer the following by writing the concept or just the page number from the text.

Identify concepts/procedures that you feel confident about:

Identify concepts/procedures that look confusing or challenging:

Be sure to ask your instructor further questions if you are still having difficulty with a concept.

Reading Time!

While **reading Section 6.6**, match the word or concept to its definition or description. Not all choices are used. *All answers are found in the Pre-Class Prep Answer Section.*

1 Perform Synthetic Division

_____ 1. Synthetic Division

_____ 2. Value of k in binomial $x - 2$

_____ 3. Top line in synthetic division

_____ 4. Bottom line in synthetic division

2 Use the Remainder Theorem to Evaluate Polynomials

_____ 5. Remainder Theorem

_____ 6. $P(3)$ is the same as...

3 Use the Factor Theorem to Factor Polynomials

_____ 7. Factor Theorem

_____ 8. Zero of the polynomial function

A. if $P(x)$ is a polynomial in x, then

$P(k) = 0$ if and only if $x - k$ is a factor of

$P(x)$

B. coefficients of the dividend

C. -2

D. if a polynomial $P(x)$ is divided by $x - k$,

the remainder is $P(k)$

E. shortcut method to divide polynomials by

a binomial of the form $x - k$

F. k, if $P(x)$ is a polynomial in x and if

$P(k) = 0$

G. coefficients of the quotient and the

remainder

H. the remainder when $P(x)$ is divided by

$x + 3$

I. 2

J. the remainder when $P(x)$ is divided by

$x - 3$

6.6 Synthetic Division: In-Class Notes

Terms, Definitions, and Main Ideas	Examples and Notes

Use notebook paper for additional notes

6.6 Synthetic Division: After Class

Important to Know

What is your homework assignment? Be sure to note it in your weekly schedule.

Section 6.6 Homework: _____ **Due**: _____

Getting to Work!

Complete your homework assignment. If you are unable to do a problem, write down the problem number and a question to help you remember what you would like to ask your instructor, your tutor, or another student.

Problem Number	Question? Where in the problem did you start to have difficulty or confusion?	Answered?

Often, you will have more questions than there is space provided here. If so, write them on notebook paper and be sure to talk to your instructor. You might ask in class or privately with the instructor.

Do You Really Know It?

Can you put into words the concepts that you learned in this section? Answer the below question from the *Writing* section in the *Study Set* in your text. Explain as if you were explaining to someone who has never taken this class before. Use notebook paper if you need more room.

Explain the factor theorem.

You Write the Test!

If you were writing the test for this section, what would you want a student to know? Write two test questions that you think might come from this material. Write questions of various difficulty; these questions can be original or chosen from the homework. Be sure to supply the answer also!

Write these questions at the end of this chapter under the section titled *My Practice Chapter 6 Test* and the answer to each question under the section titled *My Practice Chapter 6 Test Answers*.

Reflect on the Section

Look back at the *Pre-Class Prep* section. Did the lecture explain topics that you thought were going to be challenging or confusing? _____

- Are there topics that you still have questions on from the reading or the lecture? If so, complete the following:
    I don't understand...

- Speak to your instructor in class or during office hours about these concerns.

Reflect on Your Math Attitude

This section is just over half-way through this chapter. How are you doing? Think about this and list any concerns that you have as you continue your work in this chapter.

What steps, if need be, should you take to address your concerns? Talk to your instructor or tutor for suggestions.

6.7 Solving Rational Equations: Pre-Class Prep

 Are You Ready?

Complete the following problems. These review some basic skills that are needed when solving rational equations. *All answers are found in the Pre-Class Prep Answer Section.*

1. What is the LCD of $\dfrac{2}{3x}$, $\dfrac{1}{5x}$, and $\dfrac{11}{2x}$?

2. Multiply: $8(x-2)\left(\dfrac{x}{x-2}\right)$

3. Solve: $x^2 - 3x - 54 = 0$

4. Find all real numbers for which the rational expression $\dfrac{x+9}{x-4}$ is undefined.

 Reading Time!

While **reading Section 6.7**, identify the word or concept being defined. Choose from the following words (some may be used more than once or not at all). *All answers are found in the Pre-Class Prep Answer Section.*

solve a rational equation	Step 1	Step 4	extraneous solution
fraction-clearing method	Step 2	Step 5	rational equation
multiply both sides of the formula by the LCD	Step 3	Step 6	extra solution
circle the variable that you are solving for in the formula			

1 Solve Rational Equations

1. An equation that contains one or more rational expressions: _____

2. Find all values of the variable that make the equation true: _____

3. *Solving Rational Equations*

Factor all denominators: _____

Check all possible solutions in the original equation: _____

Multiply both sides of the equation by the LCD of all rational expressions in the equation:

_____

Determine which numbers cannot be solutions of the equation: _____

Use the distributive property to remove parentheses, remove any factors equal to 1, and write

the result in simplified form: _____

Solve the resulting equation: _____

2 Solve Rational Equations with Extraneous Solutions

4. A solution obtained from multiplying both sides of the equation by a quantity that contains a

variable that gives a 0 in the denominator of a rational expression: _____

3 Solve Formulas for a Specified Variable

5. Method used to solve a formula expressed as a rational equation for a specified

variable: _____

6. A success tip when solving a formula for a specified variable: _____

7. First step to solve $\dfrac{1}{R} = \dfrac{1}{R_1} + \dfrac{1}{R_2}$ for R: _____

 ## Getting Ready for Class

Briefly look through the section again. Answer the following by writing the concept or just
the page number from the text.

Identify concepts/procedures that you feel confident about:

Identify concepts/procedures that look confusing or challenging:

Be sure to ask your instructor further questions if you are still having difficulty with a concept.

Terms, Definitions, and Main Ideas	Examples and Notes

Use notebook paper for additional notes

6.7 Solving Rational Equations: After Class

Important to Know

What is your homework assignment? Be sure to note it in your weekly schedule.

Section 6.7 Homework: _____ **Due**: _____

Getting to Work!

Complete your homework assignment. If you are unable to do a problem, write down the problem number and a question to help you remember what you would like to ask your instructor, your tutor, or another student.

Problem Number	Question? Where in the problem did you start to have difficulty or confusion?	Answered?

Often, you will have more questions than there is space provided here. If so, write them on notebook paper and be sure to talk to your instructor. You might ask in class or privately with the instructor.

Do You Really Know It?

Can you put into words the concepts that you learned in this section? Answer the below question from the *Writing* section in the *Study Set* in your text. Explain as if you were explaining to someone who has never taken this class before. Use notebook paper if you need more room.

Why is it necessary to check the solutions of a rational equation?

Chapter 6 Rational Expressions and Equations

You Write the Test!

If you were writing the test for this section, what would you want a student to know? Write two test questions that you think might come from this material. Write questions of various difficulty; these questions can be original or chosen from the homework. Be sure to supply the answer also!

Write these questions at the end of this chapter under the section titled *My Practice Chapter 6 Test* and the answer to each question under the section titled *My Practice Chapter 6 Test Answers*.

Reflect on Your Math Attitude

Look back at the *Pre-Class Prep* section. Did the lecture explain topics that you thought were going to be challenging or confusing? _____

- Are there topics that you still have questions on from the reading or the lecture? If so, complete the following:
    I don't understand...

- Speak to your instructor in class or during office hours about these concerns.

Reflect on Your Math Attitude

How comfortable are you with the material presented in this section?

What could you do to either improve your comfort level or to make sure that you continue to stay comfortable with the material?

Be sure to read your math textbook to help you with the concepts presented in this section.

Are You Ready?

Complete the following problems. These review some basic skills that are needed when using rational equations to solve application problems. *All answers are found in the Pre-Class Prep Answer Section.*

1. Solve the uniform motion formula $d = rt$ for t.

2. Multiply: $\dfrac{1}{5} \cdot x$

3. Multiply: $x(x-30)\left(\dfrac{20}{x-30}\right)$

4. What is the LCD for the fractions in the rational equation $\dfrac{x}{6} + \dfrac{x}{9} = 1$?

5. Write $\dfrac{40}{9}$ as a mixed number.

6. Solve: $x^2 - 29x + 100 = 0$

 Reading Time!

(on the next page)

 Getting Ready for Class

Briefly look through the section again. Answer the following by writing the concept or just the page number from the text.

Identify concepts/procedures that you feel confident about:

Identify concepts/procedures that look confusing or challenging:

Be sure to ask your instructor further questions if you are still having difficulty with a concept.

Chapter 6 Rational Expressions and Equations

Reading Time!

While **reading Section 6.8**, match the word or concept to its definition or description. *All answers are found in the Pre-Class Prep Answer Section.*

1 Solve Shared-Work Problems

_____ 1. Shared-work problem

_____ 2. Shared-work problem formula

_____ 3. Rate of work

_____ 4. Equation that models the shared work problem with two crews

2 Solve Uniform Motion Problems

_____ 5. Uniform motion problem formula

_____ 6. Speed

_____ 7. Example of downstream travel rate

_____ 8. Example of upstream travel rate

A. (the part of the job done by 1st crew) + (the part of the job done by 2nd crew) = 1 job completed

B. another word that is often used in place of the word rate

C. Work completed = rate of work · time worked, or $W = rt$

D. problem in which two or more people (or machines) work together to complete a job

E. $12 + c$

F. for a job completed in x units of time: $\dfrac{1}{x}$

G. distance = rate · time, or $d = rt$

H. $12 - c$

6.8 Problem Solving Using Rational Equations: In-Class Notes

Terms, Definitions, and Main Ideas	Examples and Notes

Use notebook paper for additional notes

6.8 Problem Solving Using Rational Equations: After Class

Important to Know

What is your homework assignment? Be sure to note it in your weekly schedule.

Section 6.8 Homework: _____ **Due**: _____

Getting to Work!

Complete your homework assignment. If you are unable to do a problem, write down the problem number and a question to help you remember what you would like to ask your instructor, your tutor, or another student.

Problem Number	Question? Where in the problem did you start to have difficulty or confusion?	Answered?

Often, you will have more questions than there is space provided here. If so, write them on notebook paper and be sure to talk to your instructor. You might ask in class or privately with the instructor.

Do You Really Know It?

Can you put into words the concepts that you learned in this section? Answer the below question from the *Writing* section in the *Study Set* in your text. Explain as if you were explaining to someone who has never taken this class before. Use notebook paper if you need more room.

Write a shared-work problem that can be modeled by the equation: $\dfrac{x}{3}+\dfrac{x}{4}=1$.

You Write the Test!

If you were writing the test for this section, what would you want a student to know? Write two test questions that you think might come from this material. Write questions of various difficulty; these questions can be original or chosen from the homework. Be sure to supply the answer also!

Write these questions at the end of this chapter under the section titled *My Practice Chapter 6 Test* and the answer to each question under the section titled *My Practice Chapter 6 Test Answers*.

Reflect on the Section

Look back at the *Pre-Class Prep* section. Did the lecture explain topics that you thought were going to be challenging or confusing? _____

- Are there topics that you still have questions on from the reading or the lecture? If so, complete the following:
    I don't understand...

- Speak to your instructor in class or during office hours about these concerns.

Reflect on Your Math Attitude

One of the problem-solving techniques in this chapter is making a table to organize the information. Sometimes students resist making a table or drawing a diagram because it seems to be "just more work." Describe your attitude toward making tables or drawing a diagram when solving application problems.

List some benefits of making a table or drawing a diagram.

Sometimes what seems like "just more work" actually saves time in the process of solving the problem.

6.9 Proportions and Variation: Pre-Class Prep

Are You Ready?

Complete the following problems. These review some basic skills that are needed when working with ratios and proportions. *All answers are found in the Pre-Class Prep Answer Section.*

1. Multiply: a. $bd \cdot \dfrac{a}{b}$ b. $bd \cdot \dfrac{c}{d}$

2. Solve: $2(x+3) = 4(15)$

3. Solve: $x(x-7) = 2(9)$

4. Find t if $t = \dfrac{90}{n}$ and $n = 18$.

5. Let $f = 0.0036Av^2$. Find f if $A = 500$ and $v = 30$.

6. What is the degree measure of a right angle?

Reading Time!

While **reading Section 6.9**, identify the word or concept being defined. Choose from the following words (some may be used more than once). *All answers are found in the Pre-Class Prep Answer Section.*

joint variation	proportion	ratio	Step 1
the three ways to write a ratio	solving the proportion	unit cost	Step 2
constant of proportionality	extremes	means	Step 3
similar	cross products	rate	Step 4
constant of variation	combined variation	$y = kxz$ for some nonzero constant k	
The Fundamental Property of Proportions		$y = kx$ for some nonzero constant k	
units of both numerators must be the same and units of both denominators must be the same		$y = \dfrac{k}{x}$ for some nonzero constant k	

1 Identify Ratios, Rates, and Proportions

1. The quotient of 2 numbers or the quotient of 2 quantities measured with the same units: _____

2. As a fraction, using the word *to*, or with a colon: _____

3. Quotient of two quantities that have different units: _____

4. The cost per unit: _____

5. An equation indicating that two ratios or two rates are equal: _____

6. In the proportion $\dfrac{a}{b} = \dfrac{c}{d}$, the terms *a* and *d*: _____

7. In the proportion $\dfrac{a}{b} = \dfrac{c}{d}$, the terms *b* and *c*: _____

8. The product of the extremes is equal to the product of the means: _____

9. In the proportion $\dfrac{a}{b} = \dfrac{c}{d}$, the products *ad* and *bc* are called this: _____

2 Solve Proportions

10. Process of finding an unknown value in a proportion: _____

4 Solve Problems Involving Similar Triangles

11. If the three angles of the first triangle have the same measure, respectively, as the three angles of the second triangle, then the two triangles are: _____

12. If two triangles are similar, the lengths of all corresponding sides are in: _____

5 Solve Problems Involving Direct Variation

13. Equation representing '*y* varies directly as *x*' or '*y* is directly proportional to *x*': _____

14. The constant *k* in $y = kx$: _____

15. *Solving Variation Problems*

Substitute the value of *k* into the equation from step 1: _____

Translate the verbal model into the equation: _____

Substitute the remaining set of values into the equation from step 3 and solve for the unkown:

_____

Substitute the first set of values into the equation from step 1 to determine the value of *k*:

_____

Chapter 6 Rational Expressions and Equations

6 Solve Problems Involving Inverse Variation

16. Equation representing '*y* varies inversely as *x*' or '*y* is inversely proportional to *x*':_____

7 Solve Problem Involving Joint Variation

17. The relationship of 1 variable varying directly as the product of 2 or more variables: _____

18. Equation representing joint variation: _____

8 Solve Problems Involving Combined Variation

19. Applied problem involving a combination of direct and inverse variation: _____

Getting Ready for Class

     Briefly look through the section again. Answer the following by writing the concept or just the page number from the text.

Identify concepts/procedures that you feel confident about:

Identify concepts/procedures that look confusing or challenging:

Be sure to ask your instructor further questions if you are still having difficulty with a concept.

6.9 Proportions and Variation: In-Class Notes

__Terms, Definitions, and Main Ideas__	__Examples and Notes__

Use notebook paper for additional notes

6.9 Proportions and Variation: After Class

Important to Know

What is your homework assignment? Be sure to note it in your weekly schedule.

Section 6.9 Homework: _____ **Due**: _____

Getting to Work!

Complete your homework assignment. If you are unable to do a problem, write down the problem number and a question to help you remember what you would like to ask your instructor, your tutor, or another student.

Problem Number	Question? Where in the problem did you start to have difficulty or confusion?	Answered?

Often, you will have more questions than there is space provided here. If so, write them on notebook paper and be sure to talk to your instructor. You might ask in class or privately with the instructor.

Do You Really Know It?

Can you put into words the concepts that you learned in this section? Answer the below question from the *Writing* section in the *Study Set* in your text. Explain as if you were explaining to someone who has never taken this class before. Use notebook paper if you need more room.

Give examples of two quantities from everyday life that vary directly and two quantities that vary inversely.

You Write the Test!

If you were writing the test for this section, what would you want a student to know? Write two test questions that you think might come from this material. Write questions of various difficulty; these questions can be original or chosen from the homework. Be sure to supply the answer also!

Write these questions at the end of this chapter under the section titled *My Practice Chapter 6 Test* and the answer to each question under the section titled *My Practice Chapter 6 Test Answers*.

Reflect on the Section

Look back at the *Pre-Class Prep* section. Did the lecture explain topics that you thought were going to be challenging or confusing? _____

- Are there topics that you still have questions on from the reading or the lecture? If so, complete the following:
    I don't understand...

- Speak to your instructor in class or during office hours about these concerns.

Reflect on Your Math Attitude

Ratios and proportions, as with fractions, is likely a topic that you have seen before at one time or another. How did you feel when you saw that 'proportions' was the topic for this section?

Based on your attitude, list what you might do to learn the material in this section? How does reading your text fit into the goal of learning this material?

This textbook provides not only explanations but also specific examples to help you to understand the concepts.

Chapter 6 Activities

Your instructor may assign these activities to you to complete in class, or you may complete them on your own to solidify your understanding of chapter topics. The activities begin on the next page.

❖ **Student Activity:** *Tempting Expressions*

Carefully determine the steps involved in each expression and not be tempted to do incorrect mathematics by visually-pleasing expressions.

❖ **Student Activity:** *One of Us is Wrong!*

Determine which simplified version of the given complex fraction is correct and why.

❖ **Student Activity:** *Can't Use That!*

Determine what values cannot be allowed because of a "division by zero" problem, then shade the corresponding value in the provided grid.

Student Activity
Tempting Expressions

Directions: In all the expressions below, you will find yourself tempted to do incorrect mathematics by visually-pleasing expressions. Of course, just because it looks good, that doesn't mean it is the right thing to do. ☺ Think carefully about the steps involved in each expression.

1. Add: $\dfrac{5}{x} + \dfrac{x}{5}$

2. Subtract: $\dfrac{2}{x+2} - \dfrac{1}{x}$

3. Add: $\dfrac{x}{x+4} + \dfrac{x+4}{x}$

4. Divide: $\dfrac{x-1}{x+6} \div \dfrac{x+6}{x^2-36}$

5. Simplify: $\dfrac{2x+8}{2x+4}$

6. Multiply: $\dfrac{x^2+4x-12}{x^2+4x-32} \cdot \dfrac{x^2+10x+16}{x^2+10x+24}$

7. Subtract: $\dfrac{3x}{x+2} - \dfrac{x+2}{3x+3}$

8. Simplify: $\dfrac{5x^2+25x}{5x^2-125}$

9. Subtract: $\dfrac{x^2}{x^2-9} - 9$

10. Add: $\dfrac{2}{x+1} + \dfrac{2}{1-x}$

Student Activity
One of Us is Wrong!

Directions: Lou and Stu have both simplified the given complex fraction. Unfortunately, they have different answers. 1) Choose a test value for x and use it to evaluate the complex fraction and the students' answers. Give the student that is correct a ☺ for their work. 2) Then determine where the unfortunate student made the error.

1. Given expression	Lou's Work	Stu's Work
$$\dfrac{\dfrac{2x+5}{x+2}}{\dfrac{x+2}{2x+5}}$$ Test value: $x =$ ____	$$\dfrac{\dfrac{2x+5}{x+2}}{\dfrac{x+2}{2x+5}} = \dfrac{2x+5}{x+2} \div \dfrac{x+2}{2x+5}$$ $$= \dfrac{2x+5}{x+2} \cdot \dfrac{2x+5}{x+2} = \dfrac{(2x+5)^2}{(x+2)^2}$$	$$\dfrac{\dfrac{2x+5}{x+2}}{\dfrac{x+2}{2x+5}} = \dfrac{\dfrac{2x+5}{\cancel{x+2}^{\,1}}}{\dfrac{\cancel{x+2}_{\,1}}{2x+5}}$$ $$= \dfrac{\cancel{2x+5}^{\,1}}{\cancel{2x+5}_{\,1}} = 1$$
Expression value:	Expression value:	Expression value:

2. Given expression	Lou's Work	Stu's Work
$$\dfrac{2x - \dfrac{1}{2x}}{\dfrac{1}{2x} + 2x}$$ Test value: $x =$ ____	$$\dfrac{2x - \dfrac{1}{2x}}{\dfrac{1}{2x} + 2x} = \dfrac{\cancel{2x}^{\,1} - \dfrac{1}{\cancel{2x}^{\,1}}}{\dfrac{1}{\cancel{2x}^{\,1}} + \cancel{2x}^{\,1}}$$ $$= \dfrac{1-1}{1+1}$$ $$= \dfrac{0}{2} = 0$$	$$\dfrac{2x - \dfrac{1}{2x}}{\dfrac{1}{2x} + 2x} = \dfrac{\left(2x - \dfrac{1}{2x}\right)}{\left(\dfrac{1}{2x} + 2x\right)} \cdot \dfrac{\dfrac{2x}{1}}{\dfrac{2x}{1}}$$ $$= \dfrac{2x\left(\dfrac{2x}{1}\right) - \dfrac{1}{\cancel{2x}_{\,1}}\left(\dfrac{\cancel{2x}^{\,1}}{1}\right)}{\dfrac{1}{\cancel{2x}_{\,1}}\left(\dfrac{\cancel{2x}^{\,1}}{1}\right) + 2x\left(\dfrac{2x}{1}\right)} = \dfrac{4x^2 - 1}{1 + 4x^2}$$
Expression value:	Expression value:	Expression value:

3. Given expression	Lou's Work	Stu's Work
$$\dfrac{\dfrac{x^2}{2}}{\dfrac{2}{x} + \dfrac{4}{x}}$$ Test value: $x =$ ____	$$\dfrac{\dfrac{x^2}{2}}{\dfrac{2}{x} + \dfrac{4}{x}} = \dfrac{x^2}{2} \div \left(\dfrac{2}{x} + \dfrac{4}{x}\right)$$ $$= \dfrac{x^2}{2} \cdot \left(\dfrac{x}{2} + \dfrac{x}{4}\right) = \dfrac{x^2}{2} \cdot \left(\dfrac{2x}{4} + \dfrac{x}{4}\right)$$ $$= \dfrac{x^2}{2} \cdot \left(\dfrac{3x}{4}\right) = \dfrac{3x^3}{8}$$	$$\dfrac{\dfrac{x^2}{2}}{\dfrac{2}{x} + \dfrac{4}{x}} = \dfrac{\dfrac{x^2}{2}}{\dfrac{6}{x}} = \dfrac{x^2}{2} \div \dfrac{6}{x}$$ $$= \dfrac{x^2}{2} \cdot \dfrac{x}{6} = \dfrac{x^3}{12}$$
Expression value:	Expression value:	Expression value:

Cengage Student Workbook Activities, M. Andersen

Student Activity
Can't Use That!

Directions: Examine each of the equations below and decide what values cannot be allowed because of a "division by zero" problem. For the "disallowed" values, shade in the corresponding value in the grid below. The first one has been done for you.

1. $\dfrac{x+5}{x-2} = \dfrac{3}{x}; \quad x \neq 0, 2$

2. $\dfrac{x-4}{x+3} + \dfrac{2}{x-2} = 5$

3. $\dfrac{x}{3} + \dfrac{2}{x-6} = 1$

4. $\dfrac{3}{2x-1} = \dfrac{4}{x+5}$

5. $\dfrac{3}{4x} + \dfrac{1}{2} = \dfrac{2-x}{4-x}$

6. $\dfrac{5}{x+7} - \dfrac{3x}{x-10} = 6$

7. $\dfrac{x+3}{x-3} = \dfrac{4}{3x+2}$

8. $1 + \dfrac{x}{5} = \dfrac{5x}{x-8}$

9. $\dfrac{1}{4x-1} = \dfrac{4}{x}$

10. $\dfrac{9}{9+x} - 2 = \dfrac{1-x}{2x}$

11. $\dfrac{4}{x} = \dfrac{2x+1}{10}$

12. $\dfrac{x+6}{x-7} - \dfrac{9}{3x} = 4$

13. $\dfrac{x+1}{x+5} + \dfrac{2}{5x} = 3$

14. $\dfrac{x-3}{3x-4} + \dfrac{2x}{x+3} = 8$

15. $\dfrac{x}{12} + 5 = \dfrac{x+2}{x-2}$

$x \neq 0, 1, 4$	$x \neq -3, 0, 2$	$x \neq 0, 2, 4$	$x \neq -5, 0, 2$	$x \neq 0, 2$
$x \neq 0, \dfrac{1}{4}, 1$	$x \neq -3, 2, 4$	$x \neq 0, 8$	$x \neq 0, 4$	$x \neq 2, 4$
$x \neq 0, \dfrac{1}{4}$	$x \neq -3, 2$	$x \neq 8$	$x \neq 6$	$x \neq -5, 0$
$x \neq -2, 3$	$x \neq -3, -\dfrac{2}{3}, 3$	$x \neq -7, 10$	$x \neq 0, 6$	$x \neq 3, 6$
$x \neq 0$	$x \neq -\dfrac{2}{3}, 3$	$x \neq 0, 7$	$x \neq \dfrac{4}{3}, 3$	$x \neq 2$
$x \neq 0, 1, 9$	$x \neq -5, \dfrac{1}{2}$	$x \neq -6, 0$	$x \neq -7, 0, 10$	$x \neq -2, 0, 2$
$x \neq 0, 9$	$x \neq -5, 1$	$x \neq 0, -1$	$x \neq -3, 0, 3, 4$	$x \neq -1, 0, 5$

Cengage Student Workbook Activities, M. Andersen

Chapter 6 Rational Expressions and Equations

Chapter 6 Test Skills Assessment

Pre-Test Preparation Work:

1. Re-read the objectives from each section.

2. Review the *Reading Time!* activity for each section.

3. Go over all your classroom notes, if something in your notes doesn't make sense to you, make a note and ask your teacher or a classmate.

4. Make additional notations to your work if your teacher states specific concepts to study in preparation for the chapter test.

5. Attend any study sessions held by your teacher or teaching assistant.

6. Practice additional problems.

7. Go over any missed problems in your homework sets.

8. Talk out concepts with your peers in small group study sessions.

List other preparations that you have found beneficial in preparing for a math test.

Additional Practice Suggestions

1. Use your book's review problems at the back of the chapter as a practice test.

2. Take *My Practice Chapter Test* and the text's *Chapter Test*. Time yourself and do not use your notebook or textbook.

3. Pace yourself as you work through these problems.

4. Read each question carefully, playing close attention to the instructions.

5. Check your work using the answers provided.

6. Rework any missed problems. Do not just "look them over" but actually rework the problem without looking at text or notes.

My Practice Chapter 6 Test

For each section, you had the opportunity to create two test questions under the section *You Write the Test!* Write each of those questions here. Include your answers under the heading *My Practice Chapter 6 Test Answers*. **Take the test without notes or your textbook.** If you do not get a question correct, review the text and/or your notes then take the test again. For further review, do the *Chapter 6 Test* in the text.

Section 6.1

1.

2.

Section 6.2

3.

4.

Section 6.3

5.

6.

Section 6.4

7.

8.

Section 6.5

9.

10.

Section 6.6

11.

12.

Chapter 6 Rational Expressions and Equations

Section 6.7

13.

14.

Section 6.8

15.

16.

Section 6.9

17.

18.

My Practice Chapter 6 Test Answers

Section 6.1

1.

2.

Section 6.2

3.

4.

Section 6.3

5.

6.

Section 7.4

7.

8.

Section 6.5

9.

10.

Section 6.6

11.

12.

Section 6.7

13.

14.

Section 6.8

15.

16.

Section 6.9

17.

18.

Chapter 6 Rational Expressions and Equations

Chapter 7 Radical Expressions and Equations

Read the *Study Skills Workshop* found at the beginning of Chapter 7 in your textbook. **Complete** the activities below for this chapter's *Study Skills Workshop*.

Study Groups

Study groups give students an opportunity to ask their classmates' questions, share ideas, compare lecture notes, and review for tests. Would you like to be in a study group? Follow these suggestions for a successful study group.

Group Size

A study group should be small: from 3 to 6 people is best.

✓ List the students who would be in your study group. Include their contact information.

1. _____ 3. _____ 5. _____

2. _____ 4. _____ 6. _____

Time and Place

It is helpful to meet regularly in a spacious location where you can talk without disturbing others.

✓ Where will you meet? _____

✓ How often will you meet and when?_____

✓ How long will each session last? _____

Ground Rules

The group will be more effective if, early on, you agree on some rules to follow.

✓ Will you have a group leader? If so, who? _____

✓ The leader's responsibilities are _____

✓ Each session, the group will accomplish _____

✓ Prepare for each session by _____

✓ Will there be a set agenda? If so, the agenda is _____

_____

✓ When will the group discuss ways to improve the sessions? _____

_____

Are You Ready?

Complete the following problems. These review some basic skills that needed when working with square roots. *All answers are found in the Pre-Class Prep Answer Section.*

1. Evaluate: a. 15^2 b. 4^3

2. Evaluate: a. $\left(\dfrac{7}{5}\right)^2$ b. $(0.2)^2$

3. Evaluate: a. $(-6)^3$ b. 3^4

4. Simplify: a. $a^4 \cdot a^4$ b. $x^3 \cdot x^3 \cdot x^3$

5. Multiply: $(x+8)(x+8)$

6. Let $f(x) = |2x-3|$. Find $f(4)$.

Reading Time!

While **reading Section 7.1**, fill in the blanks choosing from the following words (some may be used more than once or not at all). *All answers are found in the Pre-Class Prep Answer Section.*

perfect cube	radicand	x	square root	even	rational
negative	real number(s)	index	irrational	principal	radical
cube root	radical expression	reverse	imaginary numbers	pendulum	odd
not a real number	absolute value	$\sqrt{a}$	nonnegative	$\sqrt[3]{a}$	$|x|$
$\sqrt[n]{a}$	$\sqrt{}$	$-\sqrt{a}$			

1 Find Square Roots

1. The number b is a _____ of the number a if $b^2 = a$.

2. To find the square roots of numbers_____ the squaring process.

3. A radical symbol _____ represents the positive or _____ square root of a positive number.

4. If a is a positive real number,

    a. _____ represents the positive or principal square root of a.

    b. _____ represents the negative square root of a.

    c. The _____ square root of 0 is 0: $\sqrt{0} = 0$

5. The number or variable expression within (under) a radical symbol is called the _____, and the radical symbol and radicand together are called a _____.

6. An algebraic expression containing a radical is called a _____.

7. Square roots of negative numbers such as $\sqrt{-9}$ are not _____. They are examples of_____.

8. Three important facts about square roots are

    a. If a is a perfect square, then $\sqrt{a}$, is _____.

    b. If a is a positive number that is not a perfect square, then $\sqrt{a}$ is _____.

    c. If a is a negative number then $\sqrt{a}$ is not a _____.

2 Find Square Roots of Expressions Containing Variables

9. For any real number x, $\sqrt{x^2} =$ _____, which means the _____ square root of x is equal to the _____ of x.

3 Graph the Square Root Function

10. The equation $f(x) = \sqrt{x}$ is a function called a _____ function, which belongs to a larger family of functions known as _____ functions.

11. Together, 0 and the positive real numbers are called the _____ real numbers.

4 Evaluate Radical Functions

12. The period of a _____ is the time required for the pendulum to swing back and forth to complete one cycle.

5 Find Cube Roots

13. The number b is a _____ of the number a if $b^3 = a$.

14. To find the cube roots of numbers_____ the cubing process.

Chapter 7 Radical Expressions and Equations

15. The cube root of a is written as _____. The number 3 is called the _____, a is called the _____, and the entire expression is called a _____.

16. A cube of some rational number is called a _____. The cube root of a perfect cube can be simplified: For every real number x, $\sqrt[3]{x^3} = $ _____.

6 Graph the Cube Root Function

17. The equation $f(x) = \sqrt[3]{x}$ is a function called a _____ function, which belongs to a larger family of functions known as _____ functions.

7 Find nth Roots

18. The nth root of a is written _____, where n is the _____ (or order) of the radical.

19. When n is an odd natural number, the expression $\sqrt[n]{a}$ is an _____ root, whereas, when n is an even number the expression is an _____ root.

20. To simplify $\sqrt[n]{x^n}$ if x is a real number and $n > 1$, then

   If n is an odd natural number, $\sqrt[n]{x^n} = $ _____

   If n is an even natural number, $\sqrt[n]{x^n} = $ _____

21. If n is a natural number greater than 1 and x is a real number,

   a. If $x > 0$, then $\sqrt[n]{x}$ is the positive number such that $\left(\sqrt[n]{x}\right)^n = $ _____

   b. If $x = 0$, then $\sqrt[n]{x} = $ _____

   c. If $x < 0$ $\begin{cases} \text{and } n \text{ is odd, then } \sqrt[n]{x} \text{ is the negative number such that} \left(\sqrt[n]{x}\right)^n = \underline{\quad} \\ \text{and } n \text{ is even, then } \sqrt[n]{x} \text{ is } \underline{\quad} \end{cases}$

Getting Ready for Class

Briefly look through the section again. Answer the following by writing the concept or just the page number from the text.

Identify concepts/procedures that you feel confident about:

Identify concepts/procedures that look confusing or challenging:

Be sure to ask your instructor further questions if you are still having difficulty with a concept.

Terms, Definitions, and Main Ideas	Examples and Notes

Use notebook paper for additional notes

7.1 Radical Expressions and Radical Functions: After Class

Important to Know

What is your homework assignment? Be sure to note it in your weekly schedule.

Section 7.1 Homework: _____ **Due**: _____

Getting to Work!

Complete your homework assignment. If you are unable to do a problem, write down the problem number and a question to help you remember what you would like to ask your instructor, your tutor, or another student.

Problem Number	Question? Where in the problem did you start to have difficulty or confusion?	Answered?

Often, you will have more questions than there is space provided here. If so, write them on notebook paper and be sure to talk to your instructor. You might ask in class or privately with the instructor.

Do You Really Know It?

Can you put into words the concepts that you learned in this section? Answer the below question from the *Writing* section in the *Study Set* in your text. Explain as if you were explaining to someone who has never taken this class before. Use notebook paper if you need more room.

If x is any real number, that is, if x is unrestricted, then $\sqrt{x^2} = x$ is not correct. Explain why.

Chapter 7 Radical Expressions and Equations

You Write the Test!

If you were writing the test for this section, what would you want a student to know? Write two test questions that you think might come from this material. Write questions of various difficulty; these questions can be original or chosen from the homework. Be sure to supply the answer also!

Write these questions at the end of this chapter under the section titled *My Practice Chapter 7 Test* and the answer to each question under the section titled *My Practice Chapter 7 Test Answers*.

Reflect on the Section

Look back at the *Pre-Class Prep* section. Did the lecture explain topics that you thought were going to be challenging or confusing? _____

- Are there topics that you still have questions on from the reading or the lecture? If so, complete the following:
  I don't understand…

- Speak to your instructor in class or during office hours about these concerns.

Reflect on Your Math Attitude

Does being in a study group interest you? Explain why or why not.

Being part of a study group is a good way to be actively involved in the learning of the material.

7.2 Rational Exponents: Pre-Class Prep

 Are You Ready?

Complete the following problems. These review some basic skills that are needed when working with rational exponents. *All answers are found in the Pre-Class Prep Answer Section.*

1. Evaluate: a. $\sqrt{64}$ b. $\sqrt[3]{-64}$

2. Evaluate: a. $\sqrt[4]{81}$ b. $\sqrt[5]{\dfrac{1}{32}}$

3. Simplify: a. $\sqrt{(x+5)^2}$ b. $\sqrt[3]{27a^6}$

4. Evaluate: $\left(\sqrt{36}\right)^3$

5. Simplify: 7^{-2}

6. Simplify: $\dfrac{x^5 x^7}{x^3}$

 Reading Time!

While **reading Section 7.2**, identify the word or concept being defined. Choose from the following words/expressions. (Some answers are used more than once or not at all.) *All answers are found in the Pre-Class Prep Answer Section.*

definition of $x^{1/n}$	radical form	absolute value symbols	x	Step 1
definition of $x^{-m/n}$	0	not a real number	$x < 0$	Step 2
definition of $x^{m/n}$	$x^{m/n}$	$a^4 b^3$	$x > 0$	Step 3
$a^3 b^4$	$x^{-m/n}$	$(5xyz)^{1/2}$		

Chapter 7 Radical Expressions and Equations

1 Simplify Expressions of the Form $a^{1/n}$

1. If n represents a positive integer greater than 1 and $\sqrt[n]{x}$ represents a real number, $x^{1/n} = \sqrt[n]{x}$:_____

2. If n is even and x is negative, the expression $x^{1/n}$: _____

3. *Summary of the Definitions of $x^{1/n}$ if n is a natural number greater than 1 and x is a real number.*

    a. If $x > 0$, then $x^{1/n}$ is the real number such that $\left(\sqrt[n]{x}\right)^n = ?$:_____

    b. If $x = 0$, then $x^{1/n} = ?$:_____

    c. The values of x so that $\begin{cases} \text{if } n \text{ is odd, then } x^{1/n} \text{ is the negative number such that } \left(x^{1/n}\right)^n = x \\ \text{if } n \text{ is even, then } x^{1/n} \text{ is not a real number} \end{cases}$ is true:_____

2 Simplify Expressions of the Form $a^{m/n}$

4. If m and n represent positive integers $(n \neq 1)$ and $\sqrt[n]{x}$ represents a real number, $x^{m/n} = \left(\sqrt[n]{x}\right)^m$

    and $x^{m/n} = \sqrt[n]{x^m}$:_____

3 Convert Between Radicals and Rational Exponents

5. Rules of rational exponents can be used to convert exponential form to this form:_____

6. Convert $\sqrt{5xyz}$ to exponential form:_____

4 Simplify Expressions with Negative Rational Exponents

7. If m and n are positive integers, $\dfrac{m}{n}$ is in simplified form, and $x^{m/n}$ is a real number, then

    $x^{-m/n} = \dfrac{1}{x^{m/n}}$ and $x^{m/n} = \dfrac{1}{x^{-m/n}}$ $(x \neq 0)$:_____

8. To write its reciprocal and change the sign of the exponent is another way to write this exponential expression:_____

5 Use Rules for Exponents to Simplify Expressions

9. Use the rules for exponents to simplify $\left(a^{2/3}b^{1/2}\right)^6$:_____

10. If all the variables in an expression with fractional exponents represent positive real numbers, this is not needed when the expressions are simplified: _____

6 Simplify Radical Expressions

Using Rational Exponents to Simplify Radicals

11. Simplify the rational exponents: _____

12. Change the exponential expression back into a radical: _____

13. Change the radical expression into an exponential expression: _____

Getting Ready for Class

Briefly look through the section again. Answer the following by writing the concept or just the page number from the text.

Identify concepts/procedures that you feel confident about:

Identify concepts/procedures that look confusing or challenging:

Be sure to ask your instructor further questions if you are still having difficulty with a concept.

Terms, Definitions, and Main Ideas	**Examples and Notes**

Use notebook paper for additional notes

7.2 Rational Exponents: After Class

Important to Know

What is your homework assignment? Be sure to note it in your weekly schedule.

Section 7.2 Homework: _____ **Due**: _____

Getting to Work!

Complete your homework assignment. If you are unable to do a problem, write down the problem number and a question to help you remember what you would like to ask your instructor, your tutor, or another student.

Problem Number	Question? Where in the problem did you start to have difficulty or confusion?	Answered?

Often, you will have more questions than there is space provided here. If so, write them on notebook paper and be sure to talk to your instructor. You might ask in class or privately with the instructor.

Do You Really Know It?

Can you put into words the concepts that you learned in this section? Answer the below question from the *Writing* section in the *Study Set* in your text. Explain as if you were explaining to someone who has never taken this class before. Use notebook paper if you need more room.

What is a rational exponent? Give some examples.

You Write the Test!

If you were writing the test for this section, what would you want a student to know? Write two test questions that you think might come from this material. Write questions of various difficulty; these questions can be original or chosen from the homework. Be sure to supply the answer also!

Write these questions at the end of this chapter under the section titled *My Practice Chapter 7 Test* **and the answer to each question under the section titled** *My Practice Chapter 7 Test Answers.*

Reflect on the Section

Look back at the *Pre-Class Prep* section. Did the lecture explain topics that you thought were going to be challenging or confusing? _____

- Are there topics that you still have questions on from the reading or the lecture? If so, complete the following:
    I don't understand…

- Speak to your instructor in class or during office hours about these concerns.

Reflect on Your Math Attitude

Did you notice that rational exponents are in the form of fractions? For many students, fractions have a bad name. Describe your confidence level in working with fractions.

If you are not confident, what could you do to increase your confidence?

Speak with your instructor or tutor for additional suggestions.

7.3 Simplifying and Combining Radical Expressions: Pre-Class Prep

 Are You Ready?

Complete the following problems. These review some basic skills that are needed when adding and subtracting radical expressions. *All answers are found in the Pre-Class Prep Answer Section.*

1. Complete each factorization: a. $28 = \underline{\quad} \cdot 7$ b. $54 \underline{\quad} 27 \cdot 2$

2. Simplify: a. $\sqrt[3]{-8}$ b. $\sqrt[4]{16}$

3. Multiply: $3 \cdot 3 \cdot 3 \cdot a^4 \cdot b^8$

4. Which of the following are like terms?
$$7x \quad 3x^2 \quad 9x \quad 2x^3$$

5. Combine like terms: $15m^2 + 5m - m^2 - 6m$

6. Simplify: $\dfrac{\sqrt{25}}{\sqrt{36}}$

 Reading Time!

While **reading Section 7.3**, identify the statements as True or False. *All answers are found in the Pre-Class Prep Answer Section.*

1 Use the Product Rule to Simplify Square Roots

_____ 1. The *n*th root of the product of two numbers is equal to the product of their nth roots.

_____ 2. The product rule for radicals states that if $\sqrt[n]{a}$ and $\sqrt[n]{b}$ are real numbers, $\sqrt[n]{ab} = \sqrt[n]{a}\sqrt[n]{b}$.

_____ 3. In addition to the product rule for radicals, there are rules for sums or differences.

Simplified Form of a Radical Expression:

_____ 4. Each factor in the radicand is to a power that is less than the index of the radical.

_____ 5. The radicand contains no fractions or negative numbers.

Chapter 7 Radical Expressions and Equations

_____ 6. No radicals appear in the denominator of a fraction.

2 Use Prime Factorization to Simplify Square Roots

_____ 7. Prime factorization is helpful when simplifying radical expressions.

_____ 8. The radical $3b\sqrt[3]{11b}$ is not in simplified form.

3 Use the Quotient Rule to Simplify Radical Expressions

_____ 9. For real numbers $\sqrt[n]{a}$ and $\sqrt[n]{b}$, $\sqrt[n]{\dfrac{a}{b}} = \dfrac{\sqrt[n]{a}}{\sqrt[n]{b}}$ $(b \neq 0)$.

4 Add and Subtract Radical Expressions

_____ 10. The terms $3\sqrt[4]{5}$ and $2\sqrt[3]{5}$ are like radicals.

_____ 11. If a sum or difference involves unlike radicals, simplify each one to determine whether any radicals can be combined.

_____ 12. Like radicals can be combined.

Getting Ready for Class

Briefly look through the section again. Answer the following by writing the concept or just the page number from the text.

Identify concepts/procedures that you feel confident about:

Identify concepts/procedures that look confusing or challenging:

Be sure to ask your instructor further questions if you are still having difficulty with a concept.

7.3 Simplifying and Combining Radical Expressions: In-Class Notes

Terms, Definitions, and Main Ideas	**Examples and Notes**

Use notebook paper for additional notes

Chapter 7 Radical Expressions and Equations

7.3 Simplifying and Combining Radical Expressions: After Class

Important to Know

What is your homework assignment? Be sure to note it in your weekly schedule.

Section 7.3 Homework: _____ **Due**: _____

Getting to Work!

Complete your homework assignment. If you are unable to do a problem, write down the problem number and a question to help you remember what you would like to ask your instructor, your tutor, or another student.

Problem Number	Question? Where in the problem did you start to have difficulty or confusion?	Answered?

Often, you will have more questions than there is space provided here. If so, write them on notebook paper and be sure to talk to your instructor. You might ask in class or privately with the instructor.

Do You Really Know It?

Can you put into words the concepts that you learned in this section? Answer the below question from the *Writing* section in the *Study Set* in your text. Explain as if you were explaining to someone who has never taken this class before. Use notebook paper if you need more room.

How are the procedures used to simplify $3x + 4x$ and $3\sqrt{x} + 4\sqrt{x}$ similar?

You Write the Test!

If you were writing the test for this section, what would you want a student to know? Write two test questions that you think might come from this material. Write questions of various difficulty; these questions can be original or chosen from the homework. Be sure to supply the answer also!

Write these questions at the end of this chapter under the section titled *My Practice Chapter 7 Test* and the answer to each question under the section titled *My Practice Chapter 7 Test Answers*.

Reflect on the Section

Look back at the *Pre-Class Prep* section. Did the lecture explain topics that you thought were going to be challenging or confusing? _____

- Are there topics that you still have questions on from the reading or the lecture? If so, complete the following:
  I don't understand…

- Speak to your instructor in class or during office hours about these concerns.

Reflect on Your Math Attitude

Do you know others in your class? How might your relationships with your classmates help you to be successful?

Consider this statement *"Working with others is often helpful in understanding new concepts"*. Is this true for you? Why or why not?

7.4 Multiplying and Dividing Radical Expressions: Pre-Class Prep

Are You Ready?

Complete the following problems. These review some basic skills that are needed when multiplying and dividing radical expressions. *All answers are found in the Pre-Class Prep Answer Section.*

Perform the indicated operations and simplify, if possible.

1. $\left(5a^6\right)^2$

2. $9t^3\left(6t^3 + 2t^2\right)$

3. $(2x+3)(x-1)$

4. $7 - 3\sqrt{14} + \sqrt{14} - 6$

5. $(x+10)(x-10)$

6. Build an equivalent fraction for $\dfrac{2}{3a}$ with a denominator of $27a$.

Reading Time!
(on the next page)

Getting Ready for Class

Briefly look through the section again. Answer the following by writing the concept or just the page number from the text.

Identify concepts/procedures that you feel confident about:

Identify concepts/procedures that look confusing or challenging:

Be sure to ask your instructor further questions if you are still having difficulty with a concept.

Reading Time!

While **reading Section 7.4**, match the word or concept to its definition or description. It is possible that not all choices are used. *All answers are found in the Pre-Class Prep Answer Section.*

1 Multiply Radical Expressions

_____ 1. $\sqrt[n]{a \cdot b} =$ _____

_____ 2. Product Rule for Radicals

_____ 3. Multiply radicals

_____ 4. Multiply radical expressions with more than one term

_____ 5. To simplify $3\sqrt{3}\left(4\sqrt{8} - 5\sqrt{10}\right)$

_____ 6. To simplify $\left(\sqrt{7} + \sqrt{2}\right)\left(\sqrt{7} - 9\sqrt{2}\right)$

2 Find Powers of Square Roots

_____ 7. The nth Power of the nth Root

_____ 8. $\left(2\sqrt[3]{7x^2}\right)^3 = 2^3\left(\sqrt[3]{7x^2}\right)^3$

3 Rationalize Denominators

_____ 9. Radical expression NOT simplified if…

_____ 10. Rationalizing the denominator

4 Rationalize Denominators That Have Two Terms

_____ 11. Conjugate of $\sqrt{x} - \sqrt{2}$

5 Rationalize Numerators

_____ 12. Rationalize the numerator

A. If $\sqrt[n]{a}$ is a real number, $\left(\sqrt[n]{a}\right)^n = a$

B. $\sqrt{x} + \sqrt{2}$

C. used power of a product rule for exponents

D. $\sqrt[n]{a} \cdot \sqrt[n]{b}$

E. multiply the expression by a form of 1 and use the fact that $\left(\sqrt[n]{a}\right)^n = a$ to eliminate a radical in the denominator

F. if the radicals have the same index, use the product rule for radicals

G. $\sqrt{x} - \sqrt{2}$

H. The product of the nth roots of two nonnegative numbers is equal to the nth root of the product of those numbers.

I. use the Distributive Property

J. use the same methods used to multiply polynomials with more than one term

K. multiply the numerator and denominator of the fraction by the conjugate of the numerator

L. the denominator contains a radical

M. use the FOIL method

7.4 Multiplying and Dividing Radical Expressions: In-Class Notes

Terms, Definitions, and Main Ideas	Examples and Notes

Use notebook paper for additional notes

7.4 Multiplying and Dividing Radical Expressions: After Class

Important to Know

What is your homework assignment? Be sure to note it in your weekly schedule.

Section 7.4 Homework: _____ **Due**: _____

Getting to Work!

Complete your homework assignment. If you are unable to do a problem, write down the problem number and a question to help you remember what you would like to ask your instructor, your tutor, or another student.

Problem Number	Question? Where in the problem did you start to have difficulty or confusion?	Answered?

Often, you will have more questions than there is space provided here. If so, write them on notebook paper and be sure to talk to your instructor. You might ask in class or privately with the instructor.

Do You Really Know It?

Can you put into words the concepts that you learned in this section? Answer the below question from the *Writing* section in the *Study Set* in your text. Explain as if you were explaining to someone who has never taken this class before. Use notebook paper if you need more room.

Explain why $\sqrt{m} \cdot \sqrt{m} = m$ but $\sqrt[3]{m} \cdot \sqrt[3]{m} \neq m$. Assume that m represents a positive number.

You Write the Test!

If you were writing the test for this section, what would you want a student to know? Write two test questions that you think might come from this material. Write questions of various difficulty; these questions can be original or chosen from the homework. Be sure to supply the answer also!

Write these questions at the end of this chapter under the section titled *My Practice Chapter 7 Test* and the answer to each question under the section titled *My Practice Chapter 7 Test Answers*.

Reflect on the Section

Look back at the *Pre-Class Prep* section. Did the lecture explain topics that you thought were going to be challenging or confusing? _____

- Are there topics that you still have questions on from the reading or the lecture? If so, complete the following:
    I don't understand…

- Speak to your instructor in class or during office hours about these concerns.

Reflect on Your Math Attitude

If you are in a study group, are you pleased with the time and place where you meet? Complete the following questions. If you are not in a study group, answer the questions with respect to where and when you like to study.

What I like about the time and place of the study group:

What I dislike about the time and place of the study group:

Always communicate concerns with all members of your study group. The study group needs to work well for everyone (even though sometimes compromises need to be made).

Are You Ready?

Complete the following problems. These review some basic skills that are needed when solving radical equations. *All answers are found in the Pre-Class Prep Answer Section.*

1. Simplify: a. $\left(\sqrt{x-1}\right)^2$ b. $\left(\sqrt[3]{x^3+7}\right)^3$ 2. Simplify: $\left(3\sqrt{2x+5}\right)^2$

3. Expand: $(x-4)^2$ 4. Solve: $x^2-6x-27=0$

5. Expand: $(x+2)^3$ 6. Multiply: $\left(2-\sqrt{x}\right)^2$

 Reading Time!

While **reading Section 7.5**, fill in the blanks choosing from the following words (some may be used more than once or not at all). *All answers are found in the Pre-Class Prep Answer Section.*

a	$x^n=y^n$	variable	$\left(\sqrt{a}\right)^2$	both	positive
real number	extraneous	index	nonnegative	same	solutions
radical equation	solve / solving	check	original	isolate	equal
negative	left	two	right	radical	power rule

1 Solve Equations Containing One Radical

1. A_____ is an equation that contains a variable in a radicand.

2. The Power Rule for Solving Radical Equations states: If two _____ quantities are raised to the same power, the results are _____ quantities. If x, y, and n are real numbers and $x = y$, then _____ for any exponent n.

3. The Square of a Square Root states: For any _____ real number a, _____ = _____.

4. Raising both sides of an equation to the same power can produce _____ that don't satisfy the original equation. Therefore, you must _____ each possible solution in the original equation.

5. Solutions that do not satisfy the _____ equation are called _____ solutions.

6. Strategy for _____ Radical Equations Containing Radicals:

    a. _____ a radical term on one side of the equation.

    b. Raise _____ sides of the equation to the power that is the _____ as the _____ of the radical.

    c. If it still contains a _____, go back to step a. If it does not contain a radical, _____ the resulting equation.

    d. _____ the proposed solutions in the original equation.

7. The nth Power of the nth Root: If $\sqrt[n]{a}$ is a _____, $\left(\sqrt[n]{a}\right)^n$ = _____.

2 Solve Equations Containing Two Radicals

8. To solve an equation containing _____ radicals, there must be one radical on the _____ side and one radical on the _____ side of the equation.

9. When more than one _____ appears in an equation, often the _____ must be used more than once.

3 Solve Formulas Containing Radicals

10. To solve a formula for a _____ means to _____ that variable on one side of the equation, with all other quantities on the other side.

Getting Ready for Class

Briefly look through the section again. Answer the following by writing the concept or just the page number from the text.

Identify concepts/procedures that you feel confident about:

Identify concepts/procedures that look confusing or challenging:

Be sure to ask your instructor further questions if you are still having difficulty with a concept.

Terms, Definitions, and Main Ideas	Examples and Notes

Use notebook paper for additional notes

7.5 Solving Radical Equations: After Class

Important to Know

What is your homework assignment? Be sure to note it in your weekly schedule.

Section 7.5 Homework: _____ **Due**: _____

Getting to Work!

Complete your homework assignment. If you are unable to do a problem, write down the problem number and a question to help you remember what you would like to ask your instructor, your tutor, or another student.

Problem Number	Question? Where in the problem did you start to have difficulty or confusion?	Answered?

Often, you will have more questions than there is space provided here. If so, write them on notebook paper and be sure to talk to your instructor. You might ask in class or privately with the instructor.

Do You Really Know It?

Can you put into words the concepts that you learned in this section? Answer the below question from the *Writing* section in the *Study Set* in your text. Explain as if you were explaining to someone who has never taken this class before. Use notebook paper if you need more room.

To solve the equation $\sqrt{2x+7} = \sqrt{x}$ we need only square both sides once. To solve the equation $\sqrt{2x+7} = \sqrt{x} + 2$ we have to square both sides twice. Why does the second equation require more work?

Chapter 7 Radical Expressions and Equations

You Write the Test!

If you were writing the test for this section, what would you want a student to know? Write two test questions that you think might come from this material. Write questions of various difficulty; these questions can be original or chosen from the homework. Be sure to supply the answer also!

Write these questions at the end of this chapter under the section titled *My Practice Chapter 7 Test* and the answer to each question under the section titled *My Practice Chapter 7 Test Answers*.

Reflect on the Section

Look back at the *Pre-Class Prep* section. Did the lecture explain topics that you thought were going to be challenging or confusing? _____

- Are there topics that you still have questions on from the reading or the lecture? If so, complete the following:
    I don't understand…

- Speak to your instructor in class or during office hours about these concerns.

Reflect on Your Math Attitude

Radical equations require that you check the solutions in the original equation. Have you been doing that when you work these problems? Do you find yourself willing to do this extra step? Why or why not?

Are there any other solving techniques that you have skipped throughout this course? If so, why do you think you didn't do them?

To be successful, follow the example of others who have been successful in mathematics.

7.6 Geometric Applications of Radicals: Pre-Class Prep

✓ Are You Ready?

Complete the following problems. These problems review some basic skills that are needed when working with special triangles. *All answers are found in the Pre-Class Prep Answer Section.*

1. What is the sum of the measures of the angles of any triangle?

2. Simplify: $\sqrt{3a^2}$

3. Rationalize the denominator: $\dfrac{25}{\sqrt{3}}$

4. Evaluate: $\sqrt{(9-4)^2 + (-1-11)^2}$

5. Approximate to the nearest hundredth: $2\sqrt{3}$

6. Multiply: $2 \cdot \dfrac{7\sqrt{3}}{3}$

✓ Reading Time!

(on the next page)

✓ Getting Ready for Class

Briefly look through the section again. Answer the following by writing the concept or just the page number from the text.

Identify concepts/procedures that you feel confident about:

Identify concepts/procedures that look confusing or challenging:

Be sure to ask your instructor further questions if you are still having difficulty with a concept.

Chapter 7 Radical Expressions and Equations

Reading Time!

While **reading Section 7.6**, match the word or concept to its definition or description. Not all choices are used. *All answers are found in the Pre-Class Prep Answer Section.*

1 Use the Pythagorean Theorem to Solve Problems

_____ 1. If a and b are the lengths of two legs of a right triangle and c is the length of the hypotenuse, then $a^2 + b^2 = c^2$.

_____ 2. Pythagorean equation

_____ 3. Hypotenuse

2 Solve Problems Involving $45° - 45° - 90°$ Triangles

_____ 4. Isosceles right triangle

_____ 5. Length of hypotenuse in isosceles right triangle

3 Solve Problems Involving $30° - 60° - 90°$ Triangles

_____ 6. Equilateral triangle

_____ 7. Altitude of any equilateral triangle

_____ 8. Length of hypotenuse of a $30° - 60° - 90°$ triangle

_____ 9. Length of longer leg of a $30° - 60° - 90°$ triangle

4 Use the Distance Formula to Solve Problems

_____ 10. Distance formula

A. twice as long as the shorter leg

B. triangle with three sides of equal length and three $60°$ angles

C. $d = \sqrt{\left(x_2 - x_1\right)^2 + \left(y_2 - y_1\right)^2}$

D. $\sqrt{2}$ times the length of the hypotenuse

E. right triangle with two legs of equal length

F. the side opposite the $90°$ angle

G. $\sqrt{2}$ times the length of one leg

H. divides the base into two segments of equal length and divides the equilateral triangle into two $30° - 60° - 90°$ triangles

I. $\sqrt{3}$ times the length of the shorter leg

J. $a^2 + b^2 = c^2$

K. The Pythagorean Theorem

7.6 Geometric Applications of Radicals: In-Class Notes

Terms, Definitions, and Main Ideas	Examples and Notes

Use notebook paper for additional notes

7.6 Geometric Applications of Radicals: After Class

Important to Know

What is your homework assignment? Be sure to note it in your weekly schedule.

Section 7.6 Homework: _____ **Due**: _____

Getting to Work!

Complete your homework assignment. If you are unable to do a problem, write down the problem number and a question to help you remember what you would like to ask your instructor, your tutor, or another student.

Problem Number	Question? Where in the problem did you start to have difficulty or confusion?	Answered?

Often, you will have more questions than there is space provided here. If so, write them on notebook paper and be sure to talk to your instructor. You might ask in class or privately with the instructor.

Do You Really Know It?

Can you put into words the concepts that you learned in this section? Answer the below question from the *Writing* section in the *Study Set* in your text. Explain as if you were explaining to someone who has never taken this class before. Use notebook paper if you need more room.

State the Pythagorean Theorem in words.

You Write the Test!

If you were writing the test for this section, what would you want a student to know? Write two test questions that you think might come from this material. Write questions of various difficulty; these questions can be original or chosen from the homework. Be sure to supply the answer also!

Write these questions at the end of this chapter under the section titled *My Practice Chapter 7 Test* and the answer to each question under the section titled *My Practice Chapter 7 Test Answers*.

Reflect on Your Math Attitude

Look back at the *Pre-Class Prep* section. Did the lecture explain topics that you thought were going to be challenging or confusing? _____

- Are there topics that you still have questions on from the reading or the lecture? If so, complete the following:
    I don't understand...

- Speak to your instructor in class or during office hours about these concerns.

Reflect on Your Math Attitude

Are you comfortable with geometric application of radicals problems? How would you feel if another student asked you a question about them? Do you feel confident that you could answer the question?

As you might have already heard, the best way to solve application problems successfully is to practice. Also helpful is to talk about the problems with other students. How has your study group been helpful in your understanding of these problems?

Working with other students is very helpful in that other students might bring up strategies to solve a problem that you had not thought about.

7.7 Complex Numbers: Pre-Class Prep

 Are You Ready?

Complete the following problems. These review some basic skills that are needed when working with complex numbers. *All answers are found in the Pre-Class Prep Answer Section.*

1. Explain why $\sqrt{-16}$ is not a real number.

2. Simplify: $\left(x^2 + 8x\right) + \left(5x^2 - 10x\right)$

3. Multiply: $\left(2n + 3\right)\left(7n - 1\right)$

4. Simplify: $\sqrt{63}$

5. Rationalize the denominator: $\dfrac{7}{\sqrt{x} + 4}$

6. Divide and give the remainder: $4\sqrt{87}$

 Reading Time!

While **reading Section 7.7**, identify the word or concept being defined. Choose from the following words/expressions. (Some answers are used more than once or not at all.) *All answers are found in the Pre-Class Prep Answer Section.*

$12 + 5i$	$i\sqrt{b}$	complex conjugates	-1
$3i$	$i = \sqrt{-1}$	complex number system	subtraction of complex numbers
i^R	imaginary part	multiplying complex numbers	division of complex numbers
real part	complex number	$-10 + 0i$	addition of complex numbers
adding and subtracting complex numbers			imaginary numbers

1 Express Square Roots of Negative Numbers in Terms of i

1. An expanded number system, which gives meaning to expressions such as $\sqrt{-9}$: _____

2. Definition of the imaginary number i: _____

3. Value of i^2 : _____

4. $\sqrt{-9}$ written as an imaginary number: _____

5. For any positive real number b, $\sqrt{-b} = ?$: _____

Write Complex Numbers in the Form a + bi

6. A number that can be written in the form $a + bi$, where a and b are real numbers and $i = \sqrt{-1}$:

_____

7. Complex numbers of the form $a + bi$, where $b \neq 0$: _____

8. For a complex number in the standard form $a + bi$, the real number a: _____

9. For a complex number in the standard form $a + bi$, the real number b: _____

3 Add and Subtract Complex Numbers

10. Similar to adding and subtracting polynomials: _____

11. To add the real parts and to add the imaginary parts: _____

12. To add the opposite of the complex number: _____

4 Multiply Complex Numbers

13. Similar to multiplying polynomials: _____

14. The result of $\left(-4 + 2i\right)\left(2 + i\right) = -8 - 4i + 4i + 2i^2$: _____

5 Divide Complex Numbers

15. The complex numbers $a + bi$ and $a - bi$: _____

16. To multiply the numerator and denominator by the complex conjugate of the denominator:

_____

6 Perform Operations Involving Powers of i

17. If n is a natural number that has a remainder of R when divided by 4, then $i^n = ?$: _____

Chapter 7 Radical Expressions and Equations

Getting Ready for Class

Briefly look through the section again. Answer the following by writing the concept or just the page number from the text.

Identify concepts/procedures that you feel confident about:

Identify concepts/procedures that look confusing or challenging:

Be sure to ask your instructor further questions if you are still having difficulty with a concept.

7.7 Complex Numbers: In-Class Notes

Terms, Definitions, and Main Ideas	Examples and Notes

Use notebook paper for additional notes

7.7 Complex Numbers: After Class

Important to Know

What is your homework assignment? Be sure to note it in your weekly schedule.

Section 7.7 Homework: _____ **Due**: _____

Getting to Work!

Complete your homework assignment. If you are unable to do a problem, write down the problem number and a question to help you remember what you would like to ask your instructor, your tutor, or another student.

Problem Number	Question? Where in the problem did you start to have difficulty or confusion?	Answered?

Often, you will have more questions than there is space provided here. If so, write them on notebook paper and be sure to talk to your instructor. You might ask in class or privately with the instructor.

Do You Really Know It?

Can you put into words the concepts that you learned in this section? Answer the below question from the *Writing* section in the *Study Set* in your text. Explain as if you were explaining to someone who has never taken this class before. Use notebook paper if you need more room.

The method used to divide complex numbers is similar to the method used to divide radical expressions. Explain why. Give an example.

You Write the Test!

If you were writing the test for this section, what would you want a student to know? Write two test questions that you think might come from this material. Write questions of various difficulty; these questions can be original or chosen from the homework. Be sure to supply the answer also!

Write these questions at the end of this chapter under the section titled *My Practice Chapter 7 Test* and the answer to each question under the section titled *My Practice Chapter 7 Test Answers*.

Reflect on the Section

Look back at the *Pre-Class Prep* section. Did the lecture explain topics that you thought were going to be challenging or confusing? _____

- Are there topics that you still have questions on from the reading or the lecture? If so, complete the following:
  I don't understand…

- Speak to your instructor in class or during office hours about these concerns.

Reflect on Your Math Attitude

The operations of complex numbers are said to be similar to certain operations of polynomials. Did you find this to be true? If so, what similarities or differences did you see?

Based on your observations, describe your confidence level as you work with complex numbers.

Recognizing similarities between past topics and new can help one to be more successful.

Chapter 7 Activities

Your instructor may assign these activities to you to complete in class, or you may complete them on your own to solidify your understanding of chapter topics. The activities begin on the next page.

❖ **Student Activity:** *Difficult Choices*
Determine whether absolute values are needed for the simplification of an expression, or if it is even possible to simplify.

❖ **Student Activity:** *Paint by Radicals*
Find a surprise by determining the correct simplified version of each radical expression.

❖ **Student Activity:** *Match Up on Solving Radical Equations*
Match radical equations with their solutions.

Student Activity
Difficult Choices

If the variables can represent any real number and the radical expression represents a real number:

$$\sqrt[n]{x^n} = |x| \text{ if } n \text{ is even.}$$

$$\sqrt[n]{x^n} = x \text{ if } n \text{ is odd.}$$

If the variables only represent positive values, it is not necessary to use absolute values for the even roots.

Directions: Simplify each expression and choose the simplified expression from A or B. If the radical expression represents a non-real number (an imaginary number), then choose C (non-real).

	A	B	C
1. Simplify $\sqrt{36x^2}$ if x can be any real number.	$6x$	$6|x|$	Non-real
2. Simplify $\sqrt{-36x^2}$ if x can be any real number.	$6x$	$-6x$	Non-real
3. Simplify $\sqrt{36x^2}$ if x is a positive number.	$6x$	$6|x|$	Non-real
4. Simplify $\sqrt[3]{8x^6}$ if x can be any real number.	$2x^2$	$|2x^2|$	Non-real
5. Simplify $\sqrt[3]{-8x^6}$ if x can be any real number.	$-2x^2$	$2x^2$	Non-real
6. Simplify $\sqrt{49x^2}$ if x can be any real number.	$7x$	$7|x|$	Non-real
7. Simplify $\sqrt{49x^2}$ if x is a positive number.	$7x$	$7|x|$	Non-real
8. Simplify $\sqrt{16x^4}$ if x can be any real number.	$4x^2$	$4|x^2|$	Non-real
9. Simplify $\sqrt{-16x^4}$ if x can be any real number.	$-4x^2$	$|-4x^2|$	Non-real
10. Simplify $\sqrt[3]{64x^3}$ if x can be any real number.	$4x$	$4|x|$	Non-real
11. Simplify $\sqrt[4]{16x^4}$ if x is a positive number.	$2x$	$2|x|$	Non-real
12. Simplify $\sqrt[4]{16x^4}$ if x can be any real number.	$2x$	$2|x|$	Non-real
13. Simplify $\sqrt[4]{-16x^4}$ if x is a positive number.	$2x$	$2|x|$	Non-real
14. Simplify $\sqrt[5]{-32x^5}$ if x can be any real number.	$-2x$	$2|x|$	Non-real
15. Simplify $\sqrt[5]{32x^5}$ if x can be any real number.	$2x$	$2|x|$	Non-real

Cengage Student Workbook Activities, M. Andersen

Chapter 7 Radical Expressions and Equations

Student Activity
Paint by Radicals

Directions: Simplify each expression and shade in the corresponding square in the grid below (that contains the correctly simplified version). Assume all variables represent positive numbers. The first one has been done for you. There's a surprise when you're finished!

1. Simplify $\sqrt{72} - \sqrt{50} = 6\sqrt{2} - 5\sqrt{2} = \sqrt{2}$

2. Simplify: $\sqrt{72} + \sqrt{50}$

3. Simplify: $\sqrt{16+9}$

4. Simplify: $\sqrt{40} - \sqrt{90}$

5. Simplify: $10\sqrt{20} + \sqrt{5}$

6. Simplify: $\sqrt{300} - 10\sqrt{3}$

7. Simplify: $\sqrt{75} + \sqrt{12}$

8. Simplify: $\sqrt{48y^2} - y\sqrt{27}$

9. Simplify: $\sqrt{75x^2 + 25x^2}$

10. Simplify: $\sqrt{81b^2} + \sqrt{500b}$

11. Simplify: $-13 + \sqrt{4x^2} + 5\sqrt{2} - 5\sqrt{2}$

12. Simplify: $10 + 10\sqrt[3]{5} + 2\sqrt[3]{5} + 2$

13. Simplify: $\sqrt{52b^2 - 3b^2}$

14. Simplify: $\sqrt{900} - \sqrt{80} + \sqrt{363} + \sqrt{80}$

15. Simplify: $\sqrt[3]{54} - \sqrt[3]{64} - 3\sqrt[3]{2}$

7	$21\sqrt{5}$	5	0	$11\sqrt{2}$	$10x$
$\sqrt{5}$	$y\sqrt{3}$	$\sqrt{3y}$	$2\sqrt{2xy}$	$\sqrt{122}$	$-13 + \sqrt{2} + 2x$
$11b\sqrt{5}$	$\sqrt{2}$	$\sqrt{22}$	$30 + 11\sqrt{3}$	-8	$2x - 13$
$-\sqrt{10}$	$7b$	$24\sqrt[3]{5}$	$19b\sqrt{10b}$	-4	$4 - 6\sqrt[3]{2}$
$\sqrt{-10}$	$12 + 12\sqrt[3]{5}$	$80 + 11\sqrt{3}$	$9b + 10\sqrt{5b}$	$\sqrt{3}$	$7\sqrt{3}$

Cengage Student Workbook Activities, M. Andersen

Student Activity

Match Up on Solving Radical Equations

Directions: Match each of the equations in the squares in the table below with its solution from the top. If the solution is not found among the choices A through D, then choose E (none of these).

A 5 **B** 3 **C** –2 **D** No real number solution **E** None of these

$\sqrt{7-x}=2$	$3+\sqrt{x+6}=5$	$3\sqrt{x+9}=6$
$(2y-1)^{1/2}=3$	$8+\sqrt{x-1}=5$	$6\sqrt[3]{a-4}=6$
$x=2+\sqrt{14-x}$	$\sqrt[4]{3w}=\sqrt[4]{5+2w}$	$x+\sqrt[3]{3-12x}=x+3$
$-2\sqrt{3x-1}=10$	$x-1=(7-x)^{1/2}$	$x=\sqrt{4x-11}+2$

Chapter 7 Test Skills Assessment

Pre-Test Preparation Work:

1. Re-read the objectives from each section.

2. Review the *Reading Time!* activity for each section.

3. Go over all your classroom notes, if something in your notes doesn't make sense to you, make a note and ask your teacher or a classmate.

4. Make additional notations to your work if your teacher states specific concepts to study in preparation for the chapter test.

5. Attend any study sessions held by your teacher or teaching assistant.

6. Practice additional problems.

7. Go over any missed problems in your homework sets.

8. Talk out concepts with your peers in small group study sessions.

List other preparations that you have found beneficial in preparing for a math test.

Additional Practice Suggestions

1. Use your book's review problems at the back of the chapter as a practice test.

2. Take *My Practice Chapter Test* and the text's *Chapter Test*. Time yourself and do not use your notebook or textbook.

3. Pace yourself as you work through these problems.

4. Read each question carefully, playing close attention to the instructions.

5. Check your work using the answers provided.

6. Rework any missed problems. Do not just "look them over" but actually rework the problem without looking at text or notes.

My Practice Chapter 7 Test

For each section, you had the opportunity to create two test questions under the section *You Write the Test!* Write each of those questions here. Include your answers under the heading *My Practice Chapter 7 Test Answers*. **Take the test without notes or your textbook.** If you do not get a question correct, review the text and/or your notes then take the test again. For further review, do the *Chapter 7 Test* in the text.

Section 7.1

1.

2.

Section 7.2

3.

4.

Section 7.3

5.

6.

Section 7.4

7.

8.

Section 7.5

9.

10.

Section 7.6

11.

12.

Chapter 7 Radical Expressions and Equations

Section 7.7

13.

14.

Section 7.1

1.

2.

Section 7.2

3.

4.

Section 7.3

5.

6.

Section 7.4

7.

8.

Section 7.5

9.

10.

Section 7.6

11.

12.

Section 7.7

13.

14.

Chapter 8 Quadratic Equations, Functions, and Inequalities

Read the *Study Skills Workshop* found at the beginning of Chapter 8 in your textbook. **Complete** the activities below for this chapter's *Study Skills Workshop*.

Organizing Your Notebook

In the introduction to this complete course notebook, suggestions were given to help you organize your notebook. Now is the time to give the organization of this notebook a second look. You will appreciate a well-organized notebook when it comes time to study for the final exam.

Organizing Your Notebook into Sections

Let's take a look at the organization of your notebook. Each section contains three parts: *Pre-Class Prep*, *In-Class Notes*, and *After Class*. Additional topics may be added as needed.

Organizing the Papers within Each Section

✓ *How is your organization? While each person will have their own style, check the below items that you have completed.*

_____ Write a Table of Contents to place at the beginning of your notebook.

_____ Use notebook tabs to easily access the different sections of the notebook.

_____ Place *Scheduling Your Work for This Course* pages where you can easily look at your schedule. Be sure to make copies of that form so that you can use it for each week throughout the course.

_____ Place *Peer Contact Information Guide* and *Support System Worksheet*, in addition to your study group information, in a convenient location.

_____ Divide each chapter into the three main parts mentioned above.

_____ In the *Pre-Class Prep* section, include any additional pages that you used to do the *Are You Ready?* problems and for *Getting Ready for Class*.

_____ In the *In-Class Notes* section, include all notes taken for that particular section.

_____ In the *After Class* section, *Getting to Work!* may have produced additional pages of questions and answers. Include them here for reference along with your finished homework.

_____ Through *You Write the Test!* you wrote a practice test for each chapter. This section would be a good location for returned quizzes and tests.

✓ *Compare your completed notebook with those of other students in your class. Have you overlooked any important items that would be useful when studying for the final exam?*

8.1 The Square Root Property and Completing the Square: Pre-Class Prep

Are You Ready?

Complete the following problems. These review some basic skills that are needed when using the square root property and completing the square. *All answers are found in the Pre-Class Prep Answer Section.*

1. Simplify: a. $\sqrt{28}$ b. $\sqrt{-36}$

2. Rationalize the denominator: $\sqrt{\dfrac{5}{6}}$

3. What is one-half of 9?

4. Find the square of $\dfrac{7}{2}$.

5. Fill in the blank: $1 + \dfrac{25}{144} = \dfrac{}{144}$

6. Factor: $x^2 - 8x + 16$

Reading Time!

While **reading Section 8.1**, fill in the blanks choosing from the following words (some may be used more than once or not at all). *All answers are found in the Pre-Class Prep Answer Section.*

check	constants	dividing	leading coefficient	perfect-square trinomial
multiplying	adding	binomial	square / squaring	complex numbers
$x^2 = c$	solutions	double-sign	original	square root property
one-half	coefficient	real numbers	variable	complete / completing
$x = -\sqrt{c}$	$x^2 + bx + \left(\dfrac{1}{2}b\right)^2$	$x^2 + bx - \left(\dfrac{1}{2}b\right)^2$		

1 Use the Square Root Property to Solve Quadratic Equations

1. The _____ Root Property states: For any nonnegative real number c, if _____, then $x = \sqrt{c}$ or _____.

2. The compact form for the solutions $x = \sqrt{c}$ and $x = -\sqrt{c}$ is the _____ notation: $x = \pm\sqrt{c}$

3. Some quadratic equations have solutions that are not _____.

4. The _____ can be used to solve equations that involve the square of a _____ and a constant, such as $(x-1)^2 = 16$.

2 Solve Quadratic Equations by Completing the Square

5. A quadratic equation can be solved by _____ the _____.

6. To complete the square on $x^2 + bx$, add the _____ of _____ of the coefficient of x to get

   _____.

7. *Completing the Square to Solve a Quadratic Equation in x*

   Step 1: If the _____ of x^2 is 1, go to step 2. If it is not, make it 1 by _____ both sides of the equation by the coefficient of x^2.

   Step 2: Get all _____ terms on one side of the equation and _____ on the other side.

   Step 3: _____ the square by finding _____ of the coefficient of x, _____ the result, and _____ the square to both sides of the equation.

   Step 4: Factor the _____ as the square of a binomial.

   Step 5: Solve the resulting equation using the _____.

   Step 6: _____ your answers in the _____ equation.

8. The coefficient of the squared variable of a quadratic equation is called the _____.

9. The _____ of some quadratic equations are _____ that contain i.

Getting Ready for Class

Briefly look through the section again. Answer the following by writing the concept or just the page number from the text.

Identify concepts/procedures that you feel confident about:

Identify concepts/procedures that look confusing or challenging:

Be sure to ask your instructor further questions if you are still having difficulty with a concept.

8.1 The Square Root Property and Completing the Square: In-Class Notes

Terms, Definitions, and Main Ideas	Examples and Notes

Use notebook paper for additional notes

8.1 The Square Root Property and Completing the Square: After Class

Important to Know

What is your homework assignment? Be sure to note it in your weekly schedule.

Section 8.1 Homework: _____ **Due:** _____

Getting to Work!

Complete your homework assignment. If you are unable to do a problem, write down the problem number and a question to help you remember what you would like to ask your instructor, your tutor, or another student.

Problem Number	Question? Where in the problem did you start to have difficulty or confusion?	Answered?

Often, you will have more questions than there is space provided here. If so, write them on notebook paper and be sure to talk to your instructor. You might ask in class or privately with the instructor.

Do You Really Know It?

Can you put into words the concepts that you learned in this section? Answer the below question from the *Writing* section in the *Study Set* in your text. Explain as if you were explaining to someone who has never taken this class before. Use notebook paper if you need more room.

Give an example of a perfect-square trinomial. Why do you think the word "perfect" is used to describe it?

You Write the Test!

If you were writing the test for this section, what would you want a student to know? Write two test questions that you think might come from this material. Write questions of various difficulty; these questions can be original or chosen from the homework. Be sure to supply the answer also!

Write these questions at the end of this chapter under the section titled *My Practice Chapter 8 Test* and the answer to each question under the section titled *My Practice Chapter 8 Test Answers*.

Reflect on the Section

Look back at the *Pre-Class Prep* section. Did the lecture explain topics that you thought were going to be challenging or confusing? _____

- Are there topics that you still have questions on from the reading or the lecture? If so, complete the following:
    I don't understand…

- Speak to your instructor in class or during office hours about these concerns.

Reflect on Your Math Attitude

The *Study Skills Workshop* for this chapter focused on the organization of this notebook. Think about the organization of your notebook. Are you pleased with how your notebook is organized?

What are some of the positives or negatives that you see to your organization?

Being organized is helpful in letting *you* be in control of the material from this class, not the other way around.

Are You Ready?

Complete the following problems. These review some basic skills that are needed when solving quadratic equations using the quadratic formula. *All answers are found in the Pre-Class Prep Answer Section.*

1. Evaluate: $\sqrt{5^2 - 4(4)(-6)}$

2. Simplify: $\sqrt{45}$

3. How many terms does $2x^2 - x + 7$ have? What is the coefficient of each term?

4. Evaluate: $\dfrac{-5 \pm 11}{8}$

5. Classify $\sqrt{-28}$ as a rational, irrational, or not a real number.

6. Approximate $\dfrac{6 + \sqrt{3}}{2}$ to the nearest hundredth.

Reading Time!

(on the next page)

Getting Ready for Class

Briefly look through the section again. Answer the following by writing the concept or just the page number from the text.

Identify concepts/procedures that you feel confident about:

Identify concepts/procedures that look confusing or challenging:

Be sure to ask your instructor further questions if you are still having difficulty with a concept.

Reading Time!

While **reading Section 8.2**, match the word or concept to its definition or description. Not all choices are used. *All answers are found in the Pre-Class Prep Answer Section.*

1 Derive the Quadratic Formula

_____ 1. Quadratic equation in standard form

_____ 2. Quadratic formula

2 Solve Quadratic Equations Using the Quadratic Formula

Using the Quadratic Formula:

_____ 3. Step 1

_____ 4. Step 2

_____ 5. Step 3

3 Write Equivalent Equations to Make Quadratic Formula Calculations Easier

_____ 6. To make calculations easier

4 Use the Quadratic Formula to Solve Application Problems

_____ 7. Pythagorean equation

_____ 8. Discard the negative solution

A. when a model is defined for only positive values

B. Write the equation in standard form:
$$ax^2 + bx + c = 0$$

C. $a^2 + b^2 = c^2$

D. $\dfrac{-b \pm \sqrt{b^2 - 4ac}}{2a}$, where $a \neq 0$

E. Identify $a, b,$ and c.

F. $ax^2 + bx + c = 0$, where $a > 0$

G. Substitute the values for $a, b,$ and c in the quadratic formula, $x = \dfrac{-b \pm \sqrt{b^2 - 4ac}}{2a}$, and evaluate the right side to obtain the solutions.

H. $x = \dfrac{-b \pm \sqrt{b^2 - 4ac}}{2a}$, where $a \neq 0$

I. make a simpler equivalent equation by multiplying or dividing the quadratic equation by a carefully chosen number

Chapter 8 Quadratic Equations, Functions, and Inequalities

Terms, Definitions, and Main Ideas	**Examples and Notes**

Use notebook paper for additional notes

8.2 The Quadratic Formula: After Class

Important to Know

What is your homework assignment? Be sure to note it in your weekly schedule.

Section 8.2 Homework: _____ **Due**: _____

Getting to Work!

Complete your homework assignment. If you are unable to do a problem, write down the problem number and a question to help you remember what you would like to ask your instructor, your tutor, or another student.

Problem Number	Question? Where in the problem did you start to have difficulty or confusion?	Answered?

Often, you will have more questions than there is space provided here. If so, write them on notebook paper and be sure to talk to your instructor. You might ask in class or privately with the instructor.

Do You Really Know It?

Can you put into words the concepts that you learned in this section? Answer the below question from the *Writing* section in the *Study Set* in your text. Explain as if you were explaining to someone who has never taken this class before. Use notebook paper if you need more room.

Explain why the quadratic formula, in most cases, is easier to use to solve a quadratic equation than is the method of completing the square.

You Write the Test!

If you were writing the test for this section, what would you want a student to know? Write two test questions that you think might come from this material. Write questions of various difficulty; these questions can be original or chosen from the homework. Be sure to supply the answer also!

Write these questions at the end of this chapter under the section titled *My Practice Chapter 8 Test* and the answer to each question under the section titled *My Practice Chapter 8 Test Answers*.

Reflect on the Section

Look back at the *Pre-Class Prep* section. Did the lecture explain topics that you thought were going to be challenging or confusing? _____

- Are there topics that you still have questions on from the reading or the lecture? If so, complete the following:
    I don't understand…

- Speak to your instructor in class or during office hours about these concerns.

Reflect on Your Math Attitude

The derivation of the quadratic formula was demonstrated in this section. Did you understand each step, either from your reading or an in-class presentation? If so, or if not, how did that make you feel?

Taking notes on challenging concepts provides another way of learning that information. Do you take notes during class? Have you been organizing them in this notebook in the *In-Class Notes* section?

How would taking notes and organizing them help you to be successful in this class?

8.3 The Discriminant and Equations That Can Be Written in Quadratic Form: Pre-Class Prep

✓ Are You Ready?

Complete the following problems. These review some basic concepts that are needed when working with discriminants and equations that can be written in quadratic form. *All answers are found in the Pre-Class Prep Answer Section.*

1. Find the value of the expression within the radical symbol only: $\dfrac{6 \pm \sqrt{(-6)^2 - 4(3)(-4)}}{6}$

2. Fill in the blanks: a. $x^4 = \left(x^2\right)^{\underline{}}$

                         b. $\left(x^{1/3}\right)^2 = x^{\underline{}}$

3. Fill in the blanks: a. $\left(\sqrt{x}\right)^2 = \underline{}$

                         b. $15a^{-2} = \dfrac{15}{\underline{}}$

4. Solve: $x^2 = -4$

5. Solve: a. $\sqrt{x} = 3$ b. $\sqrt[3]{x} = -\dfrac{1}{2}$

6. Solve: $\dfrac{1}{a} = \dfrac{1}{5}$

✓ Reading Time!

While **reading Section 8.3**, identify the word or concept being defined. Choose from the following words/expressions. (Some answers are used more than once or not at all.) *All answers are found in the Pre-Class Prep Answer Section.*

$\sqrt{x}$	a perfect square	positive and not a perfect square	negative
positive	completing the square	discriminant	quadratic in form
x	quadratic formula	shared-work	square root property
problem	0	factoring and zero-factor property	

1 Use the Discriminant to Determine Number and Type of Solutions

1. The expression $b^2 - 4ac$ that appears under the radical symbol in the quadratic formula:_____

   Identify the Discriminant Given Number and Type of Solution

2. Two different real numbers: _____

3. One repeated solution, a rational number: _____

4. Two different imaginary numbers that are complex conjugates: _____

5. Two different rational numbers: _____

6. Two different irrational numbers: _____

2 Solve Equations That are Quadratic in Form

   Determine best method of solution for each type of equation:

7. $ax^2 = n$ or $(ax + b)^2 = n$: _____

8. $x^2 + bx = n$, where b is even: _____

9. $ax^2 + bx + c = 0$, with easy factoring: _____

10. Any quadratic equation: _____

11. Equations that contain an expression, the same expression squared and a constant term:_____

12. An equation of the form $\left(\sqrt{x}\right)^2 - 7\sqrt{x} + 12 = 0$ is quadratic in …: _____

3 Add and Subtract Complex Numbers

13. Type of problem represented by (The fraction of tub filled with hot water) + (the fraction of the tub filled with cold water) = 1 tub filled: _____

Getting Ready for Class

Briefly look through the section again. Answer the following by writing the concept or just the page number from the text.

Identify concepts/procedures that you feel confident about:

Identify concepts/procedures that look confusing or challenging:

Be sure to ask your instructor further questions if you are still having difficulty with a concept.

8.3 The Discriminant and Equations That Can Be Written in Quadratic Form: In-Class Notes

Terms, Definitions, and Main Ideas	Examples and Notes

Use notebook paper for additional notes

8.3 The Discriminant and Equations That Can Be Written in Quadratic Form: After Class

Important to Know

What is your homework assignment? Be sure to note it in your weekly schedule.

Section 8.3 Homework: _____ **Due**: _____

Getting to Work!

Complete your homework assignment. If you are unable to do a problem, write down the problem number and a question to help you remember what you would like to ask your instructor, your tutor, or another student.

Problem Number	Question? Where in the problem did you start to have difficulty or confusion?	Answered?

Often, you will have more questions than there is space provided here. If so, write them on notebook paper and be sure to talk to your instructor. You might ask in class or privately with the instructor.

Do You Really Know It?

Can you put into words the concepts that you learned in this section? Answer the below question from the *Writing* section in the *Study Set* in your text. Explain as if you were explaining to someone who has never taken this class before. Use notebook paper if you need more room.

Describe how to predict what type of solutions the equation $3x^2 - 4x + 5 = 0$ will have.

You Write the Test!

If you were writing the test for this section, what would you want a student to know? Write two test questions that you think might come from this material. Write questions of various difficulty; these questions can be original or chosen from the homework. Be sure to supply the answer also!

Write these questions at the end of this chapter under the section titled *My Practice Chapter 8 Test* and the answer to each question under the section titled *My Practice Chapter 8 Test Answers*.

Reflect on the Section

Look back at the *Pre-Class Prep* section. Did the lecture explain topics that you thought were going to be challenging or confusing? _____

- Are there topics that you still have questions on from the reading or the lecture? If so, complete the following:
    I don't understand…

- Speak to your instructor in class or during office hours about these concerns.

Reflect on Your Math Attitude

Two sections, *Getting Ready for Class* and *Reflect on the Section*, in this notebook deal with your understanding or not understanding of concepts.

Describe how identifying concepts/procedures that you feel confident about or those that look confusing or challenging could help you to be a successful math student. Why would it be beneficial to consider whether you still have questions from the reading or lecture?

Write down your thoughts and concerns, and keep them organized so that you will be able to ask questions before quizzes or tests.

Are You Ready?

Complete the following problems. These review some basic skills that are needed when graphing quadratic functions. *All answers are found in the Pre-Class Prep Answer Section.*

1. Graph: $f(x) = x^2$

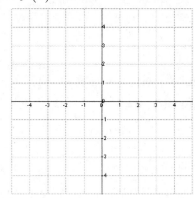

2. Let $f(x) = 2(x-3)^2 - 4$. Find $f(4)$.

3. Complete the square on $x^2 + 8x$.
   Then factor the resulting trinomial.

4. Complete the square on $x^2 + x$.
   Then factor the resulting trinomial.

5. Solve: $x^2 + 6x + 9 = 0$

6. Solve: $-2x^2 - 8x - 8 = 0$

7. Evaluate $-\dfrac{b}{2a}$ for $a = -2$ and $b = -20$.

8. Is the graph of $x = -1$ a horizontal or vertical line?

While **reading Section 8.4**, match the word or concept to its definition or description. Some choices may be used more than once or not at all. *All answers are found in the Pre-Class Prep Answer Section.*

1 Graph Functions of the Form $f(x) = ax^2$

and $f(x) = ax^2 + k$

_____ 1. Quadratic function

_____ 2. Standard form of quadratic function

_____ 3. Parabola

_____ 4. Vertex of a quadratic function

_____ 5. Axis of symmetry

_____ 6. Graph of $f(x) = ax^2$ opens down

_____ 7. Graph of $f(x) = ax^2 + k$

2 Graph Functions of the Form

$f(x) = a(x-h)^2$ *and* $f(x) = a(x-h)^2 + k$

_____ 8. Graph of $f(x) = a(x-h)^2$

_____ 9. Graph of $f(x) = a(x-h)^2 + k$

3 Graph Functions of the Form

$f(x) = ax^2 + bx + c$ *by Completing the Square*

_____ 10. Result of completing the square

4 Find the Vertex Using $-\dfrac{b}{2a}$

_____ 11. Vertex

_____ 12. The line $x = -\dfrac{b}{2a}$

_____ 13. To graph $f(x) = ax^2 + bx + c$

5 Determine Minimum and Maximum Values

_____ 14. y-coordinate of the vertex

6 Solve Quadratic Equations Graphically

_____ 15. How graph is used to find
    solutions

A. the lowest point on a parabola or the highest point on a parabola

B. $\left(-\dfrac{b}{2a}, f\left(-\dfrac{b}{2a} \right) \right)$

C. parabola with vertex at (h,k) and axis of symmetry: $x = h$

D. x-intercepts determine real-number solution(s), no x-intercept means no real-number solution

E. the axis of symmetry of the parabola

F. graph of functions of the form $f(x) = ax^2$

G. same shape as the graph of $f(x) = ax^2$ but translated k units upward if k positive, and $|k|$ units downward if k is negative

H. minimum or maximum value of the function

I. $f(x) = ax^2 + bx + c$, where a, b, c are real numbers and $a \neq 0$

J. use coefficients a, b, and c for guidance

K. when $a > 0$

L. same shape as the graph of $f(x) = ax^2$ but translated h units to the right if h positive, and $|h|$ units to the left if h is negative

M. $f(x) = a(x-h)^2 + k$

N. vertical line that passes through the vertex

O. when $a < 0$

Getting Ready for Class

Briefly look through the section again. Answer the following by writing the concept or just the page number from the text.

Identify concepts/procedures that you feel confident about:

Identify concepts/procedures that look confusing or challenging:

Be sure to ask your instructor further questions if you are still having difficulty with a concept.

8.4 Quadratic Functions and Their Graphs: In-Class Notes

Terms, Definitions, and Main Ideas	Examples and Notes

Use notebook paper for additional notes

8.4 Quadratic Functions and Their Graphs: After Class

Important to Know

What is your homework assignment? Be sure to note it in your weekly schedule.

Section 8.4 Homework: _____ **Due:** _____

Getting to Work!

Complete your homework assignment. If you are unable to do a problem, write down the problem number and a question to help you remember what you would like to ask your instructor, your tutor, or another student.

Problem Number	Question? Where in the problem did you start to have difficulty or confusion?	Answered?

Often, you will have more questions than there is space provided here. If so, write them on notebook paper and be sure to talk to your instructor. You might ask in class or privately with the instructor.

Do You Really Know It?

Can you put into words the concepts that you learned in this section? Answer the below question from the *Writing* section in the *Study Set* in your text. Explain as if you were explaining to someone who has never taken this class before. Use notebook paper if you need more room.

What are some quantities that are good to maximize? What are some quantities that are good to minimize?

You Write the Test!

If you were writing the test for this section, what would you want a student to know? Write two test questions that you think might come from this material. Write questions of various difficulty; these questions can be original or chosen from the homework. Be sure to supply the answer also!

Write these questions at the end of this chapter under the section titled *My Practice Chapter 8 Test* and the answer to each question under the section titled *My Practice Chapter 8 Test Answers*.

Reflect on the Section

Look back at the *Pre-Class Prep* section. Did the lecture explain topics that you thought were going to be challenging or confusing? _____

- Are there topics that you still have questions on from the reading or the lecture? If so, complete the following:
    I don't understand…

- Speak to your instructor in class or during office hours about these concerns.

Reflect on Your Math Attitude

What is one word that would describe your feelings about graphing: _____

How will you take that word and translate it into action on your part? What type of action do you see yourself needing to do?

Let your feelings be a guide as to what action you must take in order to be successful in mastering the material.

8.5 Quadratic and Other Nonlinear Inequalities: Pre-Class Prep

Are You Ready?

Complete the following problems. These review some basic skills that are needed when working with quadratic and nonlinear inequalities. *All answers are found in the Pre-Class Prep Answer Section.*

1. Does $x = -2$ satisfy $x^2 + x - 6 < 0$?

2. Solve: $x^2 - 5x - 50 = 0$

3. Graph the set of real numbers between -3 and 2 on a number line.

4. Graph the set of real numbers greater than 1 or less than or equal to -2 on a number line.

5. What values of x make the denominator of $\dfrac{x+1}{x^2-9}$ equal to 0?

6. Graph: $y = -x^2 + 4$

Reading Time!

While **reading Section 8.5**, fill in the blanks choosing from the following words (some may be used more than once or not at all). *All answers are found in the Pre-Class Prep Answer Section.*

original	denominator	includes / included	exclude	interval testing
boundary line	$ax^2 + bx + c > 0$	critical numbers	standard form	satisfied / not satisfied
$ax^2 + bx + c \geq 0$	allows / not allowed	quadratic	$ax^2 + bx + c \leq 0$	solid
related	test point	$ax^2 + bx + c < 0$	origin	dashed
number line	interval	test value	true	endpoints
contains	zero	linear	nonlinear	other

© 2013 Cengage Learning. All Rights Reserved. May not be scanned, copied or duplicated, or posted to a publicly accessible website, in whole or in part.

1 Solve Quadratic Inequalities

1. A _____ inequality can be written in one of the standard forms

$ax^2 + bx + c < 0$ $ax^2 + bx + c > 0$, _____, and $ax^2 + bx + c \geq 0$ where a, b, and c are real

numbers and $a \neq 0$.

2. *Solving Quadratic Inequalities*

Step 1: Write the inequality in _____ and solve its _____ quadratic equation.

Step 2: Locate the solutions (called _____) of the related quadratic equation on a _____.

Step 3: Test each _____ on the number line created in step 2 by choosing a _____ from

the interval and determining whether it satisfies the inequality. The solution set

_____ the interval(s) whose test value makes the inequality _____.

Step 4: Determine whether the _____ of the intervals are _____ in the solution set.

2 Solve Rational Inequalities

3. The _____ method can also be used to solve rational inequalities.

4. *Solving Rational Inequalities*

Step 1: Write the inequality in _____ with a single quotient on the left side and 0 on the

right side. Then solve its _____ rational equation.

Step 2: Set the _____ equal to _____ and solve that equation.

Step 3: Locate the solutions (called _____) found in steps 1 and 2 on a _____.

Step 4: Test each _____ on the number line created in step 3 by choosing a _____ from

the interval and determining whether it satisfies the inequality. The solution set _____

the interval(s) whose test value makes the inequality _____.

Step 5: Determine whether the _____ of the intervals are _____ in the solution set.

_____ any values that make the denominator 0.

3 Graph Nonlinear Inequalities in Two Variables

5. *Graphing Inequalities in Two Variables*

Step 1: Graph the related equation to find the _____ of the region. If the inequality

_____ equality (the symbol is either $\leq$ or $\geq$), draw the boundary as a _____ line.

If equality is _____ ($<$ or $>$), draw the boundary as a _____ line.

Step 2: Pick a _____ that is on one side of the boundary line. (Use the _____ if possible.) Replace x and y in the _____ inequality with the coordinates of that point. If the inequality is _____, shade the side that _____ that point. If the inequality is _____, shade the _____ side of the boundary.

6. The procedure used to graph linear inequalities in two variables is also used to graph _____ inequalities in two variables.

✓ Getting Ready for Class

Briefly look through the section again. Answer the following by writing the concept or just the page number from the text.

Identify concepts/procedures that you feel confident about:

Identify concepts/procedures that look confusing or challenging:

Be sure to ask your instructor further questions if you are still having difficulty with a concept.

8.5 Quadratic and Other Nonlinear Inequalities: In-Class Notes

<u>**Terms, Definitions, and Main Ideas**</u>

<u>**Examples and Notes**</u>

Use notebook paper for additional notes

8.5 Quadratic and Other Nonlinear Inequalities: After Class

Important to Know

What is your homework assignment? Be sure to note it in your weekly schedule.

Section 8.5 Homework: _____ **Due**: _____

Getting to Work!

Complete your homework assignment. If you are unable to do a problem, write down the problem number and a question to help you remember what you would like to ask your instructor, your tutor, or another student.

Problem Number	Question? Where in the problem did you start to have difficulty or confusion?	Answered?

Often, you will have more questions than there is space provided here. If so, write them on notebook paper and be sure to talk to your instructor. You might ask in class or privately with the instructor.

Do You Really Know It?

Can you put into words the concepts that you learned in this section? Answer the below question from the *Writing* section in the *Study Set* in your text. Explain as if you were explaining to someone who has never taken this class before. Use notebook paper if you need more room.

How are critical numbers used when solving a quadratic inequality in one variable?

You Write the Test!

If you were writing the test for this section, what would you want a student to know? Write two test questions that you think might come from this material. Write questions of various difficulty; these questions can be original or chosen from the homework. Be sure to supply the answer also!

Write these questions at the end of this chapter under the section titled *My Practice Chapter 8 Test* and the answer to each question under the section titled *My Practice Chapter 8 Test Answers*.

Reflect on the Section

Look back at the *Pre-Class Prep* section. Did the lecture explain topics that you thought were going to be challenging or confusing? _____

- Are there topics that you still have questions on from the reading or the lecture? If so, complete the following:
    I don't understand…

- Speak to your instructor in class or during office hours about these concerns.

Reflect on Your Math Attitude

How are you feeling about this chapter? How would you rate your confidence level?

List several of your concerns about this chapter.

Take a look at the various sections in your complete course notebook. Describe any action you feel you should take based on your observations.

Always take full advantage of all resources available to you as you complete this course. Past experience can help you to make positive changes for the future.

Chapter 8 Activities

 Your instructor may assign these activities to you to complete in class, or you may complete them on your own to solidify your understanding of chapter topics. The activities begin on the next page.

❖ **Student Activity:** *Square Isolation Tic-Tac-Toe*
    Play tic-tac-toe with equations easily solved with the square root property receiving an "O" in the square, otherwise an "X."

❖ **Student Activity:** *Match Up on the Discriminant*
    Classify the solution to an equation based on its discriminant.

❖ **Student Activity:** *Sink the Sub*
    Determine the y-substitution to make an equation quadratic in form.

Student Activity
Square Isolation Tic-Tac-Toe

The square root property can be used to solve a quadratic equation if
1) you can isolate the part of the equation that is written as a square.
2) the only variables in the equation are in the squared part.

Directions: If the equation **can** be solved easily with the square root property using the rules above, then isolate the squared part, and circle the equation (placing an O on the square). If the equation cannot be easily solved with the square root property, then place an X on the square.

$3(x+2)^2 - 25 = 50$	$x^2 = 5x - 4$	$-5(3x-1)^2 = -45$
$3x^2 = 75$	$24 = (x+7)^2 + 8$	$x^2 + 8x - 20 = 0$
$16x^2 - 24x = -9$	$(x-3)^2 + 2x = 0$	$x^2 + 6x = 2(3x+2)$

Student Activity
Match Up on the Discriminant

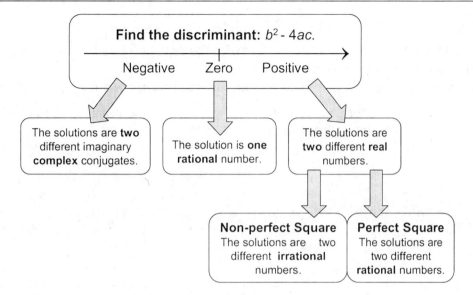

Find the discriminant: $b^2 - 4ac$.

Negative   Zero   Positive

The solutions are **two** different imaginary **complex** conjugates.

The solution is **one** **rational** number.

The solutions are **two** different **real** numbers.

Non-perfect Square
The solutions are two different **irrational** numbers.

Perfect Square
The solutions are two different **rational** numbers.

Directions: In each box, find the discriminant and then classify it.

A Two different rational numbers	**B** Two different irrational numbers	**C** One rational number	**D** Two different imaginary numbers

$25x^2 - 40x + 16 = 0$ $d = \left(-40\right)^2 - 4\left(25\right)\left(16\right) = 0$ **C**	$3x^2 + 14x - 24 = 0$	$x^2 - 2x + 2 = 0$
$9x^2 = 4$	$6x^2 - 3 = 0$	$x^2 + 3 = 5x$
$5\left(5x^2 + 2x\right) = -1$	$2x^2 + 3x = -3$	$3x^2 + 11x = 5$
$45x^2 + 24 = 67x$	$144x^2 - 264x + 121 = 0$	$4x^2 + 36x = -81$

Student Activity
Sink the Sub

Directions: For each problem, follow these steps:

- Find an appropriate y-substitution to make the equation quadratic in form.
- Solve the quadratic equation.
- Reverse the y-substitution to determine the solution set of the original equation.

The first problem is done for you. Two submarines holding the possible answers for $u = ?$ and $x = ?$ are provided to help you stay on track.

$u = ?$

$u = x^{1/3}$ $u = 2x - 3$ $u = \sqrt{x}$

$u = \sqrt{x}$ $u = \dfrac{1}{x}$ $u = x^2$

$x = ?$

$x = \{-3, 3, -i, i\}$ $x = \{-1, 4\}$

$x = \{36\}$ $x = \{-6, 5\}$

$x = \left\{64, \dfrac{1}{8}\right\}$ $x = \{25, 36\}$

Solve.	$y = ?$	Substitute, then solve for y.	Reverse the substitution	Solution set
1. $x^4 - 8x^2 - 9 = 0$	$y = x^2$	$y^2 - 8y - 9 = 0$ $(y-9)(y+1) = 0$ $y = 9, -1$	$x^2 = 9 \quad x^2 = -1$ $x = \pm 3 \quad x = \pm i$	$x = \{-3, 3, -i, i\}$
2. $x - 11\sqrt{x} + 30 = 0$				
3. $x - \sqrt{x} - 30 = 0$				
4. $2x^{2/3} - 9x^{1/3} + 4 = 0$				
5. $(2x-3)^2 = 25$				
6. $\dfrac{30}{x^2} - \dfrac{1}{x} - 1 = 0$				

Cengage Student Workbook Activities, M. Andersen

Chapter 8 Test Skills Assessment

Pre-Test Preparation Work:

1. Re-read the objectives from each section.

2. Review the *Reading Time!* activity for each section.

3. Go over all your classroom notes, if something in your notes doesn't make sense to you, make a note and ask your teacher or a classmate.

4. Make additional notations to your work if your teacher states specific concepts to study in preparation for the chapter test.

5. Attend any study sessions held by your teacher or teaching assistant.

6. Practice additional problems.

7. Go over any missed problems in your homework sets.

8. Talk out concepts with your peers in small group study sessions.

List other preparations that you have found beneficial in preparing for a math test.

Additional Practice Suggestions

1. Use your book's review problems at the back of the chapter as a practice test.

2. Take *My Practice Chapter Test* and the text's *Chapter Test*. Time yourself and do not use your notebook or textbook.

3. Pace yourself as you work through these problems.

4. Read each question carefully, playing close attention to the instructions.

5. Check your work using the answers provided.

6. Rework any missed problems. Do not just "look them over" but actually rework the problem without looking at text or notes.

My Practice Chapter 8 Test

For each section, you had the opportunity to create two test questions under the section *You Write the Test!* Write each of those questions here. Include your answers under the heading *My Practice Chapter 8 Test Answers*. **Take the test without notes or your textbook.** If you do not get a question correct, review the text and/or your notes then take the test again. For further review, do the *Chapter 8 Test* in the text.

Section 8.1

1.

2.

Section 8.2

3.

4.

Section 8.3

5.

6.

Section 8.4

7.

8.

Section 8.5

9.

10.

My Practice Chapter 8 Test Answers

Section 8.1

1.

2.

Section 8.2

3.

4.

Section 8.3

5.

6.

Section 8.4

7.

8.

Section 8.5

9.

10.

Chapter 9 Exponential and Logarithmic Functions

Read the *Study Skills Workshop* found at the beginning of Chapter 9 in your textbook. **Complete** the activities below for this chapter's *Study Skills Workshop*.

Preparing for a Final Exam

Just the thought of a final exam can send many a student into a panic. Final exams can be stressful for students because the number of topics to study can seem overwhelming. Follow the below suggestions to help reduce the stress as you prepare for the final.

Get Organized

Having all your materials where you can find them will help save time as you prepare for the final. Gather all of your notes, study sheets, homework assignments, and especially all of your returned tests to review. *You Write the Test*! in each section of this course notebook gave you the opportunity to write a test for each chapter.

✓ Complete each *My Practice Chapter Test*. Record your results and identify which chapters/sections will require additional study.

Chapter	Result	Chapters/Sections To Study
1		
2		
3		
4		
5		
6		
7		
8		
9		

Talk With Your Instructor

Knowing what the instructor deems important will help you to know what to study.

✓ Ask the instructor to list the topics that may appear on the final and those that won't be covered.

Topics that might be on the final	Topics not on the final

Manage Your Time

Studying for a final will take additional time in your schedule. Determine when, where, and what you will study each day for one week before the final.

✓ Make a detailed study plan by completing the table. You may also want to use this schedule with your study group.

When	Where	What
7 days until Final		
6 days until Final		
5 days until Final		
4 days until Final		
3 days until Final		
2 days until Final		
1 day until Final		

9.1 Algebra and Composition of Functions: Pre-Class Prep

Are You Ready?

Complete the following problems. These review some basic skills that needed when performing the algebra of functions. *All answers are found in the Pre-Class Prep Answer Section.*

1. Add: $\left(7x^2 + 5x - 12\right) + \left(2x^2 - 3x + 4\right)$

2. Subtract: $\left(8x^3 - 4x^2\right) - \left(6x^3 - 3x^2\right)$

3. Multiply: $\left(8x - 5\right)\left(9x^2 - 4\right)$

4. Divide: $\left(x^2 - x - 6\right) \div \left(x - 3\right)$

Reading Time!

While **reading Section 9.1**, identify the word or concept being defined. Choose from the following words (some may be used more than once). *All answers are found in the Pre-Class Prep Answer Section.*

$f\left(g\left(x\right)\right)$	compositions of functions	composite function
$f\left(x\right) + g\left(x\right)$	algebra of functions	$\left(C \circ F\right)\left(t\right)$
$f\left(x\right) \cdot g\left(x\right)$	$f\left(x\right) - g\left(x\right)$	functions f and g such that $h(x) = \left(f \circ g\right)(x)$
$\left(F \circ C\right)\left(t\right)$	$\dfrac{f\left(x\right)}{g\left(x\right)}$	

1 Add, Subtract, Multiply, and Divide Functions

1. The process of adding, subtracting, multiplying, and dividing functions: _____

2. Sum of functions $\left(f + g\right)\left(x\right) =:$ _____

3. Difference of functions $\left(f - g\right)\left(x\right) =:$ _____

4. Product of functions $\left(f \cdot g\right)\left(x\right) =:$ _____

5. Quotient of functions $\left(f / g\right)\left(x\right) =:$ _____

Chapter 9 Exponential and Logarithmic Functions

2 Find the Composition of Functions

6. Chains of dependence analyzed mathematically: _____

7. Example of nested parentheses: _____

8. $f \circ g$: _____

3 Use Graphs to Evaluate Functions

9. Determine the order in which function h would be evaluated for a given value of x in order to find: _____

4 Use Composite Functions to Solve Problems

10. Given the Fahrenheit temperature at time t, Celsius temperature as a function of Fahrenheit is:

_____

Getting Ready for Class

Briefly look through the section again. Answer the following by writing the concept or just the page number from the text.

Identify concepts/procedures that you feel confident about:

Identify concepts/procedures that look confusing or challenging:

Be sure to ask your instructor further questions if you are still having difficulty with a concept.

9.1 Algebra and Composition of Functions: In-Class Notes

Terms, Definitions, and Main Ideas	Examples and Notes

Use notebook paper for additional notes

405

Chapter 9 Exponential and Logarithmic Functions

9.1 Algebra and Composition of Functions: After Class

Important to Know

What is your homework assignment? Be sure to note it in your weekly schedule.

Section 9.1 Homework: _____ **Due**: _____

Getting to Work!

Complete your homework assignment. If you are unable to do a problem, write down the problem number and a question to help you remember what you would like to ask your instructor, your tutor, or another student.

Problem Number	Question? Where in the problem did you start to have difficulty or confusion?	Answered?

Often, you will have more questions than there is space provided here. If so, write them on notebook paper and be sure to talk to your instructor. You might ask in class or privately with the instructor.

Do You Really Know It?

Can you put into words the concepts that you learned in this section? Answer the below question from the *Writing* section in the *Study Set* in your text. Explain as if you were explaining to someone who has never taken this class before. Use notebook paper if you need more room.

Write out in words how to say each of the following:

$$(f \circ g)(2) \qquad\qquad g(f(-8))$$

You Write the Test!

If you were writing the test for this section, what would you want a student to know? Write two test questions that you think might come from this material. Write questions of various difficulty; these questions can be original or chosen from the homework. Be sure to supply the answer also!

Write these questions at the end of this chapter under the section titled *My Practice Chapter 9 Test* **and the answer to each question under the section titled** *My Practice Chapter 9 Test Answers.*

Reflect on the Section

Look back at the *Pre-Class Prep* section. Did the lecture explain topics that you thought were going to be challenging or confusing? _____

- Are there topics that you still have questions on from the reading or the lecture? If so, complete the following:
    I don't understand…

- Speak to your instructor in class or during office hours about these concerns.

Reflect on Your Math Attitude

Describe how you feel when you think about the final.

How have you dealt with stressful situations in the past?

Being organized can be a great stress reliever as you will know where to go for information.

Are You Ready?

Complete the following problems. These review some basic skills that are needed when working with inverse functions. *All answers are found in the Pre-Class Prep Answer Section.*

1. What are the domain and the range of the function $\{(-2,8),(3,-3),(5,10),(9,1)\}$?

2. Is a parabola that opens upward the graph of a function?

3. Fill in the blank:

   If y is a _____ of x, the symbols y and $f(x)$ are interchangeable.

4. Let $f(x) = 2x + 6$. Find $f(8x)$.

Reading Time!

While **reading Section 9.2**, fill in the blanks choosing from the following words (some may be used more than once or not at all). *All answers are found in the Pre-Class Prep Answer Section.*

vertical	once	horizontal	interchange	f^{-1}
y	one-to-one	notation	solve	x
inverse(s)	substitute	$y = x$	mirror images	(x,y)
$(f \circ f^{-1})(x)$	(y,x)	$(f^{-1} \circ f)(x)$		

1 Determine Whether a Function is a One-to-One Function

1. The function where the domain of f is the range and the range of f is the domain is called the

   _____ of f.

2. A function is called a _____ function if different inputs determine different outputs.

2 Use the Horizontal Line Test to Determine Whether a Function is One-to-One

3. The _____ line test states that a function is _____ if each horizontal line that

   intersects its graph does so exactly _____.

3 Find the Equation of the Inverse of a Function

4. If f is a _____ function consisting of ordered pairs of the form (x, y), the _____ of f,

   denoted _____, is the one-to-one function consisting of all ordered pairs of the form _____.

5. *Finding the Equation of the Inverse of a One-to-One Function*

   Step 1: If the function is written using function _____, replace $f(x)$ with _____ .

   Step 2: _____ the variables x and y.

   Step 3: _____ the resulting equation for y.

   Step 4: _____ $f^{-1}(x)$ for y.

4 Find the Composition of a Function and Its Inverse

6. The rule for the composition of _____ functions is that for any one-to-one function f and its

   inverse, f^{-1}, $(f \circ f^{-1})(x) =$ _____ and _____ $= x$.

5 Graph a Function and Its Inverse

7. Functions which are _____ of each other have graphs that are _____ about the line

   _____.

✓ Getting Ready for Class

   Briefly look through the section again. Answer the following by writing the concept or just the page number from the text.

Identify concepts/procedures that you feel confident about:

Identify concepts/procedures that look confusing or challenging:

Be sure to ask your instructor further questions if you are still having difficulty with a concept.

Terms, Definitions, and Main Ideas	Examples and Notes

Use notebook paper for additional notes

9.2 Inverse Functions: After Class

Important to Know

What is your homework assignment? Be sure to note it in your weekly schedule.

Section 9.2 Homework: _____ **Due**: _____

Getting to Work!

Complete your homework assignment. If you are unable to do a problem, write down the problem number and a question to help you remember what you would like to ask your instructor, your tutor, or another student.

Problem Number	Question? Where in the problem did you start to have difficulty or confusion?	Answered?

Often, you will have more questions than there is space provided here. If so, write them on notebook paper and be sure to talk to your instructor. You might ask in class or privately with the instructor.

Do You Really Know It?

Can you put into words the concepts that you learned in this section? Answer the below question from the *Writing* section in the *Study Set* in your text. Explain as if you were explaining to someone who has never taken this class before. Use notebook paper if you need more room.

Explain how the graph of a one-to-one function can be used to draw the graph of its inverse function.

You Write the Test!

If you were writing the test for this section, what would you want a student to know? Write two test questions that you think might come from this material. Write questions of various difficulty; these questions can be original or chosen from the homework. Be sure to supply the answer also!

Write these questions at the end of this chapter under the section titled *My Practice Chapter 9 Test* and the answer to each question under the section titled *My Practice Chapter 9 Test Answers*.

Reflect on the Section

Look back at the *Pre-Class Prep* section. Did the lecture explain topics that you thought were going to be challenging or confusing? _____

- Are there topics that you still have questions on from the reading or the lecture? If so, complete the following:
  I don't understand…

- Speak to your instructor in class or during office hours about these concerns.

Reflect on Your Math Attitude

This course notebook gives the opportunity for you to create your own practice test for each chapter in the *You Write the Test!* section. Have you created your own practice tests? If so, how did that make you feel? If not, why?

Compare your own practice tests to the actual tests that have been given. Did you make good predictions? Write down your thoughts about this comparison.

Learning to look critically at chapter topics will help you to determine what is on the test.

Are You Ready?

Complete the following problems. These review some basic skills that are needed when working with exponential functions. *All answers are found in the Pre-Class Prep Answer Section.*

1. Simplify: a. $3^4 \cdot 3^8$ b. $\left(3^4\right)^8$

2. Evaluate: a. 2^3 b. 2^0 c. 2^{-2}

3. Evaluate: a. $\left(\dfrac{1}{3}\right)^2$ b. $\left(\dfrac{1}{3}\right)^0$ c. $\left(\dfrac{1}{3}\right)^{-3}$

4. Fill in the blanks: The graph of $f(x) = x$ is a _____ and the graph of $f(x) = x^2$ is a _____.

 Reading Time!
(on the next page)

 Getting Ready for Class

Briefly look through the section again. Answer the following by writing the concept or just the page number from the text.

Identify concepts/procedures that you feel confident about:

Identify concepts/procedures that look confusing or challenging:

Be sure to ask your instructor further questions if you are still having difficulty with a concept.

413   Chapter 9 Exponential and Logarithmic Functions

Reading Time!

While **reading Section 9.3**, match the word or concept to its definition or description. Not all choices are used. *All answers are found in the Pre-Class Prep Answer Section.*

1 Define Exponential Functions

_____ 1. Exponential function

_____ 2. Domain of $f(x) = b^x$

_____ 3. Range of $f(x) = b^x$

2 Graph Exponential Functions

_____ 4. Point-plotting method

_____ 5. Graph of exponential growth

_____ 6. Horizontal asymptote of $f(x) = b^x$

_____ 7. Graph of exponential decay

_____ 8. y-intercept of $f(x) = b^x$

_____ 9. Graph of $f(x) = b^x$ passes through

this point

_____ 10. Exponential functions are this type

of function

_____ 11. $f(x) = b^x$ is an increasing function

if

_____ 12. $f(x) = b^x$ is an decreasing

function if

3 Use Exponential Functions if Applications Involving Growth or Decay

_____ 13. Compound interest formula

_____ 14 . Simple interest formula

A. $0 < b < 1$

B. $I = Prt$

C. $(0, \infty)$

D. $f(x) = b^x$ or $y = b^x$ where $b > 0, b \neq 1$,

and x is a real number

E. $(1, b)$

F.

G. $(0, 1)$

H. $(-\infty, \infty)$

I. x-axis

J. create a table of function values and plot

the corresponding ordered pairs

K.

L. one-to-one

M. $b > 1$

N. $A = P\left(1 + \dfrac{r}{k}\right)^{kt}$

9.3 Exponential Functions: In-Class Notes

Terms, Definitions, and Main Ideas	Examples and Notes

Use notebook paper for additional notes

9.3 Exponential Functions: After Class

Important to Know

What is your homework assignment? Be sure to note it in your weekly schedule.

Section 9.3 Homework: _____ **Due:** _____

Getting to Work!

Complete your homework assignment. If you are unable to do a problem, write down the problem number and a question to help you remember what you would like to ask your instructor, your tutor, or another student.

Problem Number	Question? Where in the problem did you start to have difficulty or confusion?	Answered?

Often, you will have more questions than there is space provided here. If so, write them on notebook paper and be sure to talk to your instructor. You might ask in class or privately with the instructor.

Do You Really Know It?

Can you put into words the concepts that you learned in this section? Answer the below question from the *Writing* section in the *Study Set* in your text. Explain as if you were explaining to someone who has never taken this class before. Use notebook paper if you need more room.

How do the graphs of $f(x) = 3^x$ and $g(x) = \left(\frac{1}{3}\right)^x$ differ? How are they similar?

You Write the Test!

If you were writing the test for this section, what would you want a student to know? Write two test questions that you think might come from this material. Write questions of various difficulty; these questions can be original or chosen from the homework. Be sure to supply the answer also!

Write these questions at the end of this chapter under the section titled *My Practice Chapter 9 Test* and the answer to each question under the section titled *My Practice Chapter 9 Test Answers*.

 ## Reflect on the Section

Look back at the *Pre-Class Prep* section. Did the lecture explain topics that you thought were going to be challenging or confusing? _____

- Are there topics that you still have questions on from the reading or the lecture? If so, complete the following:
    I don't understand…

- Speak to your instructor in class or during office hours about these concerns.

 ## Reflect on Your Math Attitude

In this section, you were able to graph exponential functions. How did you feel when you saw that you would be graphing again?

Describe how your previous experiences with graphing helped you (or didn't) with the topic of this section.

Often concepts in mathematics appear over and over again, typically with further depth or application. Make sure that you understand concepts so that your foundation is strong.

© 2013 Cengage Learning. All Rights Reserved. May not be scanned, copied or duplicated, or posted to a publicly accessible website, in whole or in part.

9.4 Logarithmic Functions: Pre-Class Prep

 Are You Ready?

Complete the following problems. These review some basic skills that are needed when working with logarithmic functions. *All answers are found in the Pre-Class Prep Answer Section.*

1. A table of values for a one-to-one function f is shown below. Complete the table of values for f^{-1}.

x	$f(x)$
0	1
1	2
2	4

x	$f^{-1}(x)$
1	
2	
4	

2. Fill in the blanks: a. $5^{-3} = \dfrac{1}{\underline{\hspace{1cm}}}$ b. $3^{} = 27$ 3. Fill in the blanks: a. $4^{} = 4$ b. $7^{} = \sqrt{7}$

4. Evaluate: a. 10^3 b. 10^{-2}

 Reading Time!

While **reading Section 9.4**, identify the statements as True or False. *All answers are found in the Pre-Class Prep Answer Section.*

1 Define Logarithm

_____ 1. Definition of logarithm: For all positive numbers b, where $b \neq 1$, and all positive numbers x, $y = \log_b x$ is equivalent to $x = b^y$.

_____ 2. An exponent is a logarithm.

2 Write Logarithmic Equations as Exponential Equations

_____ 3. $\log_7 \sqrt{7} = \dfrac{1}{2}$ means $7^{\frac{1}{2}} = \sqrt{7}$.

_____ 4. $\log_4 64$ can be read as $\log_4 \cdot 64$.

3 Write Exponential Equations as Logarithmic Equations

_____ 5. $x = b^y$ is equivalent to $x = \log_b y$.

_____ 6. Certain logarithmic equations can be solved by writing them as exponential equations.

4 Evaluate Logarithmic Expressions

_____ 7. $\log_b x$ is the exponent to which b is raised to get x, in symbols $b^{\log_b x} = x$.

_____ 8. Base-10 logarithms are also called natural logarithms.

_____ 9. $\log x$ means $\log_{10} x$, so that $\log 10^x = x$.

5 Graph Logarithmic Functions

_____ 10. If $b > 0$ and $b \neq 1$, the logarithmic function with base b is defined by the equations
$f(x) = \log_b x$ or $y = \log_b x$.

_____ 11. The domain of $f(x) = \log_b x$ is the interval $(0, \infty)$ and the range is the interval $(-\infty, \infty)$.

_____ 12. The graph of $f(x) = \log_b x$ passes through the point $(1, b)$.

_____ 13. The graph of $f(x) = \log_b x$ passes through the point $(0, 1)$.

_____ 14. For the graph of $f(x) = \log_b x$, the y-axis is an asymptote.

6 Use Logarithmic Formulas and functions in Applications

_____ 15. The formula for decibel voltage gain is $\text{dB gain} = 20 \log \dfrac{E_0}{E_1}$.

✓ Getting Ready for Class

Briefly look through the section again. Answer the following by writing the concept or just the page number from the text.

Identify concepts/procedures that you feel confident about:

Identify concepts/procedures that look confusing or challenging:

Be sure to ask your instructor further questions if you are still having difficulty with a concept.

Chapter 9 Exponential and Logarithmic Functions

9.4 Logarithmic Functions: In-Class Notes

Terms, Definitions, and Main Ideas	Examples and Notes

Use notebook paper for additional notes

9.4 Logarithmic Functions: After Class

Important to Know

What is your homework assignment? Be sure to note it in your weekly schedule.

Section 9.4 Homework: _____ **Due**: _____

Getting to Work!

Complete your homework assignment. If you are unable to do a problem, write down the problem number and a question to help you remember what you would like to ask your instructor, your tutor, or another student.

Problem Number	Question? Where in the problem did you start to have difficulty or confusion?	Answered?

Often, you will have more questions than there is space provided here. If so, write them on notebook paper and be sure to talk to your instructor. You might ask in class or privately with the instructor.

Do You Really Know It?

Can you put into words the concepts that you learned in this section? Answer the below question from the *Writing* section in the *Study Set* in your text. Explain as if you were explaining to someone who has never taken this class before. Use notebook paper if you need more room.

Explain why it is impossible to find the logarithm of a negative number.

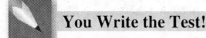

You Write the Test!

If you were writing the test for this section, what would you want a student to know? Write two test questions that you think might come from this material. Write questions of various difficulty; these questions can be original or chosen from the homework. Be sure to supply the answer also!

Write these questions at the end of this chapter under the section titled *My Practice Chapter 9 Test* and the answer to each question under the section titled *My Practice Chapter 9 Test Answers*.

Reflect on the Section

Look back at the *Pre-Class Prep* section. Did the lecture explain topics that you thought were going to be challenging or confusing? _____

- Are there topics that you still have questions on from the reading or the lecture? If so, complete the following:
  I don't understand…

- Speak to your instructor in class or during office hours about these concerns.

Reflect on Your Math Attitude

Think about the amount of time that you spend studying for this class. How has your attitude affected the amount of time that you have spent studying?

When planning for the final, do you feel that you would spend some time each day reviewing and preparing for the final? Why or why not?

Starting to review early will not only break the material into manageable amounts to study but will also give you time to ask questions of your instructor, tutor, study group, or another student.

9.5 Base-*e* Exponential and Logarithmic Functions: Pre-Class Prep

 Are You Ready?

Complete the following problems. These review some basic skills that are needed when working with base-*e* exponential and logarithmic functions. *All answers are found in the Pre-Class Prep Answer Section.*

1. Evaluate $\left(1+\dfrac{1}{n}\right)^{n}$ for $n = 2$.
2. Evaluate: $(2.718)^{0}$

3. Round 2.718281828459 to the nearest tenth.

4. Use a calculator to evaluate $55(2.718)^{0.4}$. Round to the nearest hundredth.

5. Fill in the blank: $\log_{3} x = 6$ is equivalent to _____ = _____.

6. Fill in the blank: If $(3, 8)$ is on the graph of a one-to-one function f, then the point $(\underline{\quad}, \underline{\quad})$ is on the graph of f^{-1}.

 Reading Time!

While **reading Section 9.5**, fill in the blanks choosing from the following words (some may be used more than once or not at all). *All answers are found in the Pre-Class Prep Answer Section.*

natural	semiannually	increase	*x*-axis	range	decrease
horizontal	domain	double	$(0, 1)$	$A = Pe^{rt}$	continuously
$(1, e)$	*y*-axis	quarterly	translated	exponential	exponent
$(-\infty, \infty)$	$\ln x$	power	logarithmic	$e^{\ln x}$	$f(x) = \ln x$
$y = \ln x$	$(0, \infty)$	$\dfrac{\ln 2}{r}$	annual growth rate		

Chapter 9 Exponential and Logarithmic Functions

1 Define the Natural Exponential Function

1. The function defined by $f(x) = e^x$ is the _____ exponential function where $e = 2.71828...$ The _____ of $f(x) = e^x$ is the interval $(-\infty, \infty)$. The _____ is the interval $(0, \infty)$.

2 Graph the Natural Exponential Function

2. The graph of $f(x) = e^x$ passes the _____ line test, so the function is one-to-one. The _____ is the horizontal asymptote, the y-intercept is _____, there is no x-intercept, and the graph passes through the point _____.

3. The graph of the natural exponential function can be_____ horizontally and vertically.

3 Use Base-e Exponential Formulas and Functions in Applications

4. The formula for exponential growth/decay is _____, where r is called the _____. If r is negative, the growth represents a _____.

5. Interest paid two times per year is said to be compounded _____, four times per year is compounded _____, and infinitely many times a year is compounded _____.

4 Define Base-e Logarithms

6. Base-e logarithms are called _____ logarithms, where _____ is written rather than $\log_e x$.

7. The logarithm of a number is an _____, so for natural logarithms, _____ is the exponent to which e is raised to get x, in symbols _____ $= x$.

5 Evaluate Natural Logarithmic Expressions

8. To evaluate $\ln e$, ask "To what _____ must we raise e to get e?"

9. Certain natural _____ equations can be solved by writing them as equivalent natural _____ equations.

6 Graph the Natural Logarithmic Function

10. The natural logarithmic function with base e is defined by the equations _____ or _____, where $\ln x = \log_e x$.

11. The domain of $f(x) = \ln x$ is the interval _____ and the range is the interval _____.

7 Use Base-e Formulas and Functions in Applications

12. The doubling time is the time required for a population to _____ at a certain annual rate.

13. If r is the annual rate, compounded _____, and t is the time required for a population to double, then $t =$ _____.

✓ Getting Ready for Class

Briefly look through the section again. Answer the following by writing the concept or just the page number from the text.

Identify concepts/procedures that you feel confident about:

Identify concepts/procedures that look confusing or challenging:

Be sure to ask your instructor further questions if you are still having difficulty with a concept.

9.5 Base-*e* Exponential and Logarithmic Functions: In-Class Notes

Terms, Definitions, and Main Ideas	Examples and Notes

Use notebook paper for additional notes

9.5 Base-e Exponential and Logarithmic Functions: After Class

Important to Know

What is your homework assignment? Be sure to note it in your weekly schedule.

Section 9.5 Homework: _____ **Due**: _____

Getting to Work!

Complete your homework assignment. If you are unable to do a problem, write down the problem number and a question to help you remember what you would like to ask your instructor, your tutor, or another student.

Problem Number	Question? Where in the problem did you start to have difficulty or confusion?	Answered?

Often, you will have more questions than there is space provided here. If so, write them on notebook paper and be sure to talk to your instructor. You might ask in class or privately with the instructor.

Do You Really Know It?

Can you put into words the concepts that you learned in this section? Answer the below question from the *Writing* section in the *Study Set* in your text. Explain as if you were explaining to someone who has never taken this class before. Use notebook paper if you need more room.

Explain the difference between the functions $f(x) = \log x$ and $g(x) = \ln x$.

You Write the Test!

If you were writing the test for this section, what would you want a student to know? Write two test questions that you think might come from this material. Write questions of various difficulty; these questions can be original or chosen from the homework. Be sure to supply the answer also!

Write these questions at the end of this chapter under the section titled *My Practice Chapter 9 Test* and the answer to each question under the section titled *My Practice Chapter 9 Test Answers*.

Reflect on the Section

Look back at the *Pre-Class Prep* section. Did the lecture explain topics that you thought were going to be challenging or confusing? _____

- Are there topics that you still have questions on from the reading or the lecture? If so, complete the following:
    I don't understand...

- Speak to your instructor in class or during office hours about these concerns.

Reflect on Your Math Attitude

How confident do you feel about being able to determine what type of questions might be on the final?

Discuss with your instructor about the potential topics on the final. Do you feel comfortable talking with your instructor? Why or why not?

Talking with your instructor could provide insight into what is expected on the final.

9.6 Properties of Logarithms: Pre-Class Prep

 Are You Ready?

Complete the following problems. These review some basic skills that are needed when working with properties of logarithms. *All answers are found in the Pre-Class Prep Answer Section.*

1. Evaluate: $\log_2 8 + \log_2 1$

2. Evaluate: $\log 10,000 - \log 10$

3. Evaluate: $9\log_3 \dfrac{1}{3}$

4. Evaluate: $\dfrac{\log_7 49}{\log_7 7}$

5. a. Write $\sqrt{x}$ using a fractional exponent.

6. Use a calculator to find $\ln 5$. Round to four decimal places.

   b. Write $(x-2)^{\frac{1}{2}}$ using radical notation.

 Reading Time!

While **reading Section 9.6**, identify the word or concept being defined. Choose from the following words. *All answers are found in the Pre-Class Prep Answer Section.*

$p\log_b M$	$b^{\log_b x} = x \ (x > 0)$	$\log_b b^x = x$	hydrogen ion
$\log_b M - \log_b N$	$\log_b 1 = 0$	change-of-base formula	$-\log\left[H^+\right]$
$\log_b b = 1$	$\log_b M + \log_b N$	condense logarithmic expressions	
$\dfrac{\log_a x}{\log_a b}$	pH scale	expand a logarithmic expression	

1 Use the Four Basic Properties of Logarithms

1. First property of logarithms: _____

2. Second property of logarithms: _____

3. Third property of logarithms: _____

4. Fourth property of logarithms: _____

2 Use the Product Rule for Logarithms

5. For all positive real numbers, M, N, and b, where $b \neq 1$, $\log_b MN = ?$: _____

3 Use the Quotient Rule for Logarithms

6. For all positive real numbers, M, N, and b, where $b \neq 1$, $\log_b \dfrac{M}{N} = ?$: _____

4 Use the Power Rule for Logarithms

7. For all positive real numbers, M, N, b, where $b \neq 1$, and any real number p, $\log_b M^p = ?$: _____

8. To write a logarithmic expression as the sum and/or difference of logarithms of a single quantity using the properties of logarithms: _____

5 Write Logarithmic Expressions as a Single Logarithm

9. To write a logarithmic expression as 1 logarithm using the properties of logarithms in reverse: _____

6 Use the Change-of-Base Formula

10. To find a logarithm with some base other than 10 or e: _____

11. For any logarithmic bases a and b, and any positive real number x, $\log_b x = ?$: _____

7 Use Properties of Logarithms to Solve Application Problems

12. Positively charged hydrogen atom missing its electron: _____

13. Measures the concentration of hydrogen ions in a solution: _____

14. If $\left[H^+ \right]$ is the hydrogen ion concentration in gram-ions per liter, then pH = ?: _____

✓ Getting Ready for Class

Briefly look through the section again. Answer the following by writing the concept or just the page number from the text.

Identify concepts/procedures that you feel confident about:

Identify concepts/procedures that look confusing or challenging:

Be sure to ask your instructor further questions if you are still having difficulty with a concept.

Terms, Definitions, and Main Ideas	Examples and Notes

Use notebook paper for additional notes

9.6 Properties of Logarithms: After Class

Important to Know

What is your homework assignment? Be sure to note it in your weekly schedule.

Section 9.6 Homework: _____ Due: _____

Getting to Work!

Complete your homework assignment. If you are unable to do a problem, write down the problem number and a question to help you remember what you would like to ask your instructor, your tutor, or another student.

Problem Number	Question? Where in the problem did you start to have difficulty or confusion?	Answered?

Often, you will have more questions than there is space provided here. If so, write them on notebook paper and be sure to talk to your instructor. You might ask in class or privately with the instructor.

Do You Really Know It?

Can you put into words the concepts that you learned in this section? Answer the below question from the *Writing* section in the *Study Set* in your text. Explain as if you were explaining to someone who has never taken this class before. Use notebook paper if you need more room.

Explain the difference between a logarithm of a product and the product of logarithms.

You Write the Test!

If you were writing the test for this section, what would you want a student to know? Write two test questions that you think might come from this material. Write questions of various difficulty; these questions can be original or chosen from the homework. Be sure to supply the answer also!

Write these questions at the end of this chapter under the section titled *My Practice Chapter 9 Test* and the answer to each question under the section titled *My Practice Chapter 9 Test Answers*.

Reflect on the Section

Look back at the *Pre-Class Prep* section. Did the lecture explain topics that you thought were going to be challenging or confusing? _____

- Are there topics that you still have questions on from the reading or the lecture? If so, complete the following:
    I don't understand…

- Speak to your instructor in class or during office hours about these concerns.

Reflect on Your Math Attitude

What did you think about this chapter? Do you feel confident in your ability to work these problems?

How might you improve or maintain your confidence level as this chapter comes to a close?

The questions that you wrote for *My Practice Chapter 9 Test* in the *You Write the Test!* section will provide an opportunity to review some of the concepts in this chapter.

9.7 Exponential and Logarithmic Equations: Pre-Class Prep

 ## Are You Ready?

Complete the following problems. These review some basic skills that are needed when solving exponential and logarithmic equations. *All answers are found in the Pre-Class Prep Answer Section.*

1. Fill in the blanks: a. $16 = 2^-$

 b. $\dfrac{1}{125} = 5^-$

2. Find $\dfrac{\log 12}{\log 9}$. Round to four decimal places.

3. Use the power rule for logarithms: $\log_3 8^x$

4. Evaluate: $\ln e$

5. Use a property of logarithms to write each expression as a single logarithm:
   a. $\log_2 5 + \log_2 x$
   b. $\ln 10 - \ln(2t + 1)$

6. Evaluate: $\log(-6)$

 ## Reading Time!
(on the next page)

 ## Getting Ready for Class

Briefly look through the section again. Answer the following by writing the concept or just the page number from the text.

Identify concepts/procedures that you feel confident about:

Identify concepts/procedures that look confusing or challenging:

Be sure to ask your instructor further questions if you are still having difficulty with a concept.

Reading Time!

While **reading Section 9.7**, match the word or concept to its definition or description. Not all choices are used. *All answers are found in the Pre-Class Prep Answer Section.*

1 Solve Exponential Equations

_____ 1. Exponential equation

_____ 2. Exponent property of equality

_____ 3. Logarithm property of equality

Strategy for Solving Exponential

Equations

_____ 4. Step 1

_____ 5. Step 2

_____ 6. Step 3

_____ 7. Step 4

2 Solve Logarithmic Equations

_____ 8. Logarithmic equation

_____ 9. Extraneous solutions

3 Use Exponential and Logarithmic Equations to Solve Application Problems

_____ 10. pH of a solution

_____ 11. Half-life

_____ 12. Radioactive decay formula

_____ 13. Exponential growth model

A. If step 2 is difficult or impossible, take the common or natural logarithm on both sides. Use the power rule of logarithms to write the variable exponent as a factor, and then solve the resulting equation.

B. equation with a logarithmic expression that contains a variable

C. For any positive number b, where $b \neq 1$, and positive numbers x and y, $\log_b x = \log_b y$ is equivalent to $x = y$.

D. equation that contains a variable in one of its exponents

E. $\mathrm{pH} = -\log\left[\mathrm{H}^+\right]$

F. If both sides of the equation can be written as exponential expressions with the same base, do so. Then set the exponents equal and solve the resulting equation.

G. possible solutions that produce the logarithm of a negative number or the logarithm of 0 in the original equation

H. For any real number b, where $b \neq -1, 0$, or 1, $b^x = b^y$ is equal to $x = y$.

I. $P = P_0 e^{kt}$

J. Isolate one of the exponential expressions in the equation.

K. Check the results in the original equation.

L. $A = A_0 2^{-t/h}$

M. time it takes for half of a sample of a radioactive material to decompose

9.7 Exponential and Logarithmic Equations: In-Class Notes

Terms, Definitions, and Main Ideas	Examples and Notes

Use notebook paper for additional notes

9.7 Exponential and Logarithmic Equations: After Class

Important to Know

What is your homework assignment? Be sure to note it in your weekly schedule.

Section 9.7 Homework: _____ **Due**: _____

Getting to Work!

Complete your homework assignment. If you are unable to do a problem, write down the problem number and a question to help you remember what you would like to ask your instructor, your tutor, or another student.

Problem Number	Question? Where in the problem did you start to have difficulty or confusion?	Answered?

Often, you will have more questions than there is space provided here. If so, write them on notebook paper and be sure to talk to your instructor. You might ask in class or privately with the instructor.

Do You Really Know It?

Can you put into words the concepts that you learned in this section? Answer the below question from the *Writing* section in the *Study Set* in your text. Explain as if you were explaining to someone who has never taken this class before. Use notebook paper if you need more room.

Explain how to solve the equation $2^{x+1} = 32$.

You Write the Test!

If you were writing the test for this section, what would you want a student to know? Write two test questions that you think might come from this material. Write questions of various difficulty; these questions can be original or chosen from the homework. Be sure to supply the answer also!

Write these questions at the end of this chapter under the section titled *My Practice Chapter 9 Test* **and the answer to each question under the section titled** *My Practice Chapter 9 Test Answers.*

Reflect on the Section

Look back at the *Pre-Class Prep* section. Did the lecture explain topics that you thought were going to be challenging or confusing? _____

- Are there topics that you still have questions on from the reading or the lecture? If so, complete the following:
    I don't understand…

- Speak to your instructor in class or during office hours about these concerns.

Reflect on Your Math Attitude

Logarithmic equations require that you check the solutions in the original equation. Have you been doing that when you work these problems? Do you find yourself willing to do this extra step? Why or why not?

Are there any other solving techniques that you have not done in this chapter? If so, why do you think you didn't do them?

As you prepare for the final or the chapter test, identify solving strategies that must be used to be successful.

Chapter 9 Exponential and Logarithmic Functions

Chapter 9 Activities

Your instructor may assign these activities to you to complete in class, or you may complete them on your own to solidify your understanding of chapter topics. The activities begin on the next page.

❖ **Student Activity:** *Exponential Workout*

Simplify exponential expressions to find their equivalent form in a grid.

❖ **Student Activity:** *Paint by Equivalent Exponents*

Find the picture by converting between logarithmic equations and exponential equations.

❖ **Student Activity:** *Escape the Logjam*

Solve exponential or logarithmic equations and simplify exponential or logarithmic expressions to determine your movement through the logjam.

Student Activity
Exponential Workout

Directions: For each expression, simplify and then find and shade the box with its equivalent form in the grid.

1. $\left(e^{\sqrt{2}}\right)^{\sqrt{2}}$

2. $\dfrac{2}{\sqrt{2}}$

3. $\dfrac{10^{3\sqrt{4}}}{10^{2\sqrt{4}}}$

4. e^{-1}

5. $\dfrac{6^7}{6^3 \cdot 36}$

6. $\left(\dfrac{1}{3}\right)^{-1}$

7. $\left(e^2\right)^{3/2}$

8. $\left(\sqrt{13}\right)^4$

9. $7^{36}7^{-34}$

10. $3460\left(\dfrac{1}{50}\right)^2$

11. $\dfrac{1}{\left(|-2|\right)^{-99}}$

12. $1-\dfrac{e^0}{9}$

13. $\left(\dfrac{1}{e}\right)^{-1}$

14. $\left(5^{2\sqrt{5}}\right)^3$

15. $\left(9^4\right)^{1/2}$

16. $0.001\left(10^3\right)$

17. $10^x \cdot 10^{3x}$

18. $\left(4e\right)^{1/2}$

19. $\dfrac{6^{\sqrt{3}+2\sqrt{2}}}{6^{\sqrt{8}}}$

20. $2\left(256\right)^{-1}2^8$

21. $3^x - \dfrac{1}{3^{-x}}$

22. $\dfrac{e^2}{e^{-3}}$

1	e	$2\sqrt{e}$	$2^{2\sqrt{2}}$	4^{4x}		14	99	36		2
3	-2	$9^{\sqrt{2}}$		e^2	27^{12}	49		$6^{\sqrt{3}}$	34	e^3
	$5^{6\sqrt{5}}$	0	$\dfrac{1}{2}$	5	$\dfrac{8}{9}$	$\dfrac{1}{4}$	$\dfrac{2}{3}$	81	1.384	
2^{99}	10			$\dfrac{1}{e}$	$\dfrac{6^4}{36}$	10^{4x}		169	3^{2-x}	9
$\sqrt{2}$	100	8^x	9		$-\dfrac{1}{3}$		-4	e^5	$2^{2\sqrt{2}}$	

Cengage Student Workbook Activities, M. Andersen

Chapter 9 Exponential and Logarithmic Functions

Student Activity
Paint by Equivalent Equations

Directions: Write each logarithmic equation as its corresponding exponential equation. Write each exponential equation as its corresponding logarithmic equation. Shade in your answers in the grid to reveal the picture. The first one has been done for you.

1. $5^3 = 125$

$\log_5 125 = 3$

2. $2^x = 8$

3. $\log x = 2$

4. $\log_4 64 = 3$

5. $10^x = \dfrac{1}{100}$

6. $16^x = 2$

7. $\log_3 x = -1$

8. $\log 1 = 0$

9. $4^{-x} = 2$

10. $\left(\dfrac{1}{2}\right)^x = 4$

11. $x^2 = 9$

12. $\log 1000 = x$

13. $\log_x 25 = 2$

14. $x^0 = 1$

15. $\left(\dfrac{1}{9}\right)^x = 3$

$\log_x 9 = 2$	$\log_2 9 = x$	$\log_x 3 = \dfrac{1}{9}$	$\log_5 125 = 3$	$\log_1 x = 0$	$\log \dfrac{1}{100} = x$
$10^2 = x$	$2^{10} = x$	$x = \log_8 2$	$\log_2 4 = -x$	$\log_{1/2} 4 = x$	$\log x = \dfrac{1}{100}$
$\log_{1/9} 3 = x$	$\log_x 1 = 0$	$\log_2 8 = x$	$\log_4 2 = -x$	$\log_4 x = 4$	$10^0 = 1$
$3^{-1} = x$	$x^{-1} = 3$	$4^3 = 64$	$3^4 = 81$	$\log_4 \dfrac{1}{2} = x$	$\log_3 x = \dfrac{1}{9}$
$10^x = 1000$	$x^2 = 25$	$\log_{16} 2 = x$	$\log_2 16 = x$	$25^x = 2$	$2^x = 25$

Student Activity
Escape the Logjam

Directions: Begin at the box marked START. Either solve the equation or simplify the expression. Look for the result in the movement grid to the right. If your result is not in the movement grid, that's a BAD sign. For example, the answer to $3^x = 7$ is $x \approx 1.7712$, so we move to the right. Next solve $\log(3-x) = 3$.

Move Left	Move Right	Move Down	
0.0025	~~1.7712~~	3.4190	1.0986
23.1331	0.4474	0.7906	8
	2.4022	2	−2
	0.2091	−997	10
		4.3774	

START $\;\rightarrow \rightarrow \rightarrow$ $3^x = 7$	$\log(3-x) = 3$	$e^0 + e^x$
$4^x = 16$	$4^{x-2} = 27$	$\log_5 25 = x$
$e^{-2.4t} = 14.2$	$e^{2.8x} = 3.5$	$\log_2(x-7) + \log_2 x = 3$
$e^{3x} = 60$	$\log 8 - \log x = 2$	$2^{x+2} = 3^x$
$3^{x^2 - 3x + 4} = 81$	$2^x = 16$	$\log 3x = \log 6$
$\log 5 - \log 2x = 0.5$	$5^x = 7^{x-4}$	$\log \dfrac{2x}{5} = -3$
$e^{2x} = 9$	$\ln x = 5$	$\log_2 5x - \log_2 3 = 4$
$3^x - 11 = 3$	$5^{x+1} = 7$	$\log x^{900} = 900$
$\dfrac{1}{3}\log(3x+5) = \log x$	$\ln 5x = 4$	$6^{x+4} = 36$
$3^{9y-3} = 1$	$z + \ln e^{-z}$	**ESCAPE** the logjam

Chapter 9 Test Skills Assessment

Pre-Test Preparation Work:

1. Re-read the objectives from each section.

2. Review the *Reading Time!* activity for each section.

3. Go over all your classroom notes, if something in your notes doesn't make sense to you, make a note and ask your teacher or a classmate.

4. Make additional notations to your work if your teacher states specific concepts to study in preparation for the chapter test.

5. Attend any study sessions held by your teacher or teaching assistant.

6. Practice additional problems.

7. Go over any missed problems in your homework sets.

8. Talk out concepts with your peers in small group study sessions.

List other preparations that you have found beneficial in preparing for a math test.

Additional Practice Suggestions

1. Use your book's review problems at the back of the chapter as a practice test.

2. Take *My Practice Chapter Test* and the text's *Chapter Test*. Time yourself and do not use your notebook or textbook.

3. Pace yourself as you work through these problems.

4. Read each question carefully, playing close attention to the instructions.

5. Check your work using the answers provided.

6. Rework any missed problems. Do not just "look them over" but actually rework the problem without looking at text or notes.

My Practice Chapter 9 Test

For each section, you had the opportunity to create two test questions under the section *You Write the Test!* Write each of those questions here. Include your answers under the heading *My Practice Chapter 9 Test Answers*. **Take the test without notes or your textbook.** If you do not get a question correct, review the text and/or your notes then take the test again. For further review, do the *Chapter 9 Test* in the text.

Section 9.1

1.

2.

Section 9.2

3.

4.

Section 9.3

5.

6.

Section 9.4

7.

8.

Section 9.5

9.

10.

Section 9.6

11.

12.

Chapter 9 Exponential and Logarithmic Functions

Section 9.7

13.

14.

My Practice Chapter 9 Test Answers

Section 9.1

1.

2.

Section 9.2

3.

4.

Section 9.3

5.

6.

Section 9.4

7.

8.

Section 9.5

9.

10.

Section 9.6

11.

12.

Section 9.7

13.

14.

Chapter 10 Conic Sections; More Graphing

Read the *Study Skills Workshop* found at the beginning of Chapter 10 in your textbook. **Complete** the activities below for this chapter's *Study Skills Workshop*.

Preparing for Your Next Math Course

Before moving on to a new mathematics course, it is worthwhile to take some time to reflect on your effort and performance in this course.

✓ *Answer the following questions.*

How was my attendance?

Was I organized? Did I have the right materials?

Did I follow a regular schedule?

Did I pay attention in class and take good notes?

Did I spend the appropriate amount of time on homework?

How did I prepare for tests? Did I have a test-taking strategy?

Was I part of a study group? If no, why not? If so, was it worthwhile?

Did I ever seek extra help from a tutor or from my instructor?

In what topics was I the strongest? In what topics was I the weakest?

If I had it to do over, would I do anything differently?

Are You Ready?

Complete the following problems. These review some basic skills that are needed when working with circles and parabolas. *All answers are found in the Pre-Class Prep Answer Section.*

1. Use a special-product formula to find $(x-8)^2$.

2. Complete the square on $x^2 - 6x$ and factor the resulting trinomial.

3. Factor: $y^2 + 10y + 25$

4. Factor out -3 from the terms of the expression $-3y^2 - 12y$.

Reading Time!

While **reading Section 10.1**, fill in the blanks choosing from the following words (some may be used more than once or not at all). *All answers are found in the Pre-Class Prep Answer Section.*

focus	$x = h$	parabolas	distance	radius
ellipses	hyperbolas	points	intersection	equidistant
paraboloid	$(x-h)^2 + (y-k)^2 = r^2$	$y = a(x-h)^2 + k$	$a < 0$	standard
(h, k)	$x^2 + y^2 + Dx + Ey + F = 0$	$a > 0$	directrix	$x = ay^2 + by + c$
circles	$y = ax^2 + bx + c$	$x = a(y-k)^2 + h$	focus	center

$$d = \sqrt{(x_2 - x_1)^2 + (y_2 - y_1)^2}$$

1 Identify Conic Sections and Some of Their Applications

1. Conic sections are curves formed by the _____ of a plane with an infinite right-circular cone.

   There are four basic shapes, called _____, _____, _____, and _____.

2. Any light or sound placed at the _____ of a _____, which is a dish-shaped surface, is reflected outward in parallel lines.

2 Graph Equations of Circles Written in Standard Form

3. A circle is the set of all _____ in a plane that are a fixed distance from a fixed point called its _____. The fixed distance is called the _____ of the circle.

4. The _____ form of the equation of a circle with radius r and center at _____ is
   _____.

3 Write the Equation of a Circle, Given Its Center and Radius

5. The equation of a circle can be written by knowing its _____ and _____.

4 Convert the General Form of the Equation of a Circle to Standard Form

6. The general form of the equation of a circle is _____.

5 Solve Application Problems Involving Circles

7. To find the distance between two points, use the _____ formula _____.

6 Convert the General Form of the Equation of a Parabola to Standard Form to Graph It.

8. A parabola is the set of all points in a plane that are _____ from a fixed point, called the _____, and a fixed line, called the _____.

9. The graph of the quadratic function _____, where $a \neq 0$ is a parabola with vertex at _____. The axis of symmetry is the line_____ . The parabola opens upward when _____ and downward when _____.

10. The general forms of the equation of a parabola are _____ and _____.

11. The standard form for the equation of a parabola that opens to the right or left is _____, where $a > 0$ if the parabola opens to the right and $a < 0$ if the parabola opens to the left.

Getting Ready for Class

Briefly look through the section again. Answer the following by writing the concept or just the page number from the text.

Identify concepts/procedures that you feel confident about:

Identify concepts/procedures that look confusing or challenging:

Be sure to ask your instructor further questions if you are still having difficulty with a concept.

10.1 The Circle and the Parabola: In-Class Notes

Terms, Definitions, and Main Ideas	Examples and Notes

Use notebook paper for additional notes

10.1 The Circle and the Parabola: After Class

Important to Know

What is your homework assignment? Be sure to note it in your weekly schedule.

Section 10.1 Homework: _____ **Due**: _____

Getting to Work!

Complete your homework assignment. If you are unable to do a problem, write down the problem number and a question to help you remember what you would like to ask your instructor, your tutor, or another student.

Problem Number	Question? Where in the problem did you start to have difficulty or confusion?	Answered?

Often, you will have more questions than there is space provided here. If so, write them on notebook paper and be sure to talk to your instructor. You might ask in class or privately with the instructor.

Do You Really Know It?

Can you put into words the concepts that you learned in this section? Answer the below question from the *Writing* section in the *Study Set* in your text. Explain as if you were explaining to someone who has never taken this class before. Use notebook paper if you need more room.

From the equation of a circle, explain how to determine the radius and the coordinates of the center.

You Write the Test!

If you were writing the test for this section, what would you want a student to know? Write two test questions that you think might come from this material. Write questions of various difficulty; these questions can be original or chosen from the homework. Be sure to supply the answer also!

Write these questions at the end of this chapter under the section titled *My Practice Chapter 10 Test* **and the answer to each question under the section titled** *My Practice Chapter 10 Test Answers.*

Reflect on the Section

Look back at the *Pre-Class Prep* section. Did the lecture explain topics that you thought were going to be challenging or confusing? _____

- Are there topics that you still have questions on from the reading or the lecture? If so, complete the following:
    I don't understand…

- Speak to your instructor in class or during office hours about these concerns.

Reflect on Your Math Attitude

Write one word that describes how you feel about taking another math course.

_____

Was the word you wrote a positive word, a negative word, or a neutral word?

What happened in this math course that influenced your choice of word?

Past experience can help you to make positive changes for the future.

10.2 The Ellipse: Pre-Class Prep

 Are You Ready?

Complete the following problems. These review some basic skills that are needed when working with ellipses. *All answers are found in the Pre-Class Prep Answer Section.*

1. Solve: $a^2 = 81$

2. Simplify: $y - (-4)$

3. Multiply: $16\left(\dfrac{x^2}{4} + \dfrac{y^2}{16}\right)$

4. Simplify: $\sqrt{8}$

 Reading Time!
(on the next page)

 Getting Ready for Class

Briefly look through the section again. Answer the following by writing the concept or just the page number from the text.

Identify concepts/procedures that you feel confident about:

Identify concepts/procedures that look confusing or challenging:

Be sure to ask your instructor further questions if you are still having difficulty with a concept.

Reading Time!

While **reading Section 10.2**, match the word or concept to its definition or description. Not all choices are used. *All answers are found in the Pre-Class Prep Answer Section.*

1 Define an Ellipse

_____ 1. Definition of an ellipse

2 Graph Ellipses Centered at the Origin

_____ 2. Equation of an ellipse in standard form

_____ 3. *x*-intercepts of an ellipse

_____ 4. *y*-intercepts of an ellipse

_____ 5. Major axis

_____ 6. Center

_____ 7. Minor axis

3 Graph Ellipses Centered at (h, k)

_____ 8. The standard form of the equation of ellipse centered at (h, k)

4 Solve Application Problems Involving Ellipses

_____ 9. Reflective properties

A. light or sound originating at one focus of an ellipse is reflected by the interior of the figure to the other focus

B. line segment joining the vertices

C. $(a,0)$ and $(-a,0)$

D. set of all points in a plane for which the sum of the distances from two fixed points is a constant

E. line segment whose endpoints are on the ellipse and is perpendicular to the major axis at the center

F. $(0,b)$ and $(0,-b)$

G. midpoint of the major axis

H. $\dfrac{x^2}{a^2}+\dfrac{y^2}{b^2}=1$, where $a>0$ and $b>0$

I. $\dfrac{(x-h)^2}{a^2}+\dfrac{(y-k)^2}{b^2}=1$, where $a>0$ and $b>0$

Chapter 10 Conic Sections; More Graphing

10.2 The Ellipse: In-Class Notes

Terms, Definitions, and Main Ideas	Examples and Notes

Use notebook paper for additional notes.

10.2 The Ellipse: After Class

Important to Know

What is your homework assignment? Be sure to note it in your weekly schedule.

Section 10.2 Homework: _____ **Due**: _____

Getting to Work!

Complete your homework assignment. If you are unable to do a problem, write down the problem number and a question to help you remember what you would like to ask your instructor, your tutor, or another student.

Problem Number	Question? Where in the problem did you start to have difficulty or confusion?	Answered?

Often, you will have more questions than there is space provided here. If so, write them on notebook paper and be sure to talk to your instructor. You might ask in class or privately with the instructor.

Do You Really Know It?

Can you put into words the concepts that you learned in this section? Answer the below question from the *Writing* section in the *Study Set* in your text. Explain as if you were explaining to someone who has never taken this class before. Use notebook paper if you need more room.

Explain the difference between the focus of an ellipse and the vertex of an ellipse.

You Write the Test!

If you were writing the test for this section, what would you want a student to know? Write two test questions that you think might come from this material. Write questions of various difficulty; these questions can be original or chosen from the homework. Be sure to supply the answer also!

Write these questions at the end of this chapter under the section titled *My Practice Chapter 10 Test* and the answer to each question under the section titled *My Practice Chapter 10 Test Answers*.

Reflect on the Section

Look back at the *Pre-Class Prep* section. Did the lecture explain topics that you thought were going to be challenging or confusing? _____

- Are there topics that you still have questions on from the reading or the lecture? If so, complete the following:
    I don't understand…

- Speak to your instructor in class or during office hours about these concerns.

Reflect on Your Math Attitude

Look back at your overall attitude in this course. Do you feel that your attitude helped or hurt your performance?

What do you think you could do in order to make your attitude either continue in a positive way or to improve your attitude when you take another math course?

Should you need to improve your attitude, speak with your instructor or a counselor for suggestions.

10.3 The Hyperbola: Pre-Class Prep

Are You Ready?

Complete the following problems. These review some basic concepts that are needed when working with hyperbolas. *All answers are found in the Pre-Class Prep Answer Section.*

1. Solve: $b^2 = 16$

2. Simplify: $x - (-2)$

3. What is the slope of the line represented by $y = \frac{2}{3}x$?

4. If $xy = 20$, find y when $x = -2$.

Reading Time!

While **reading Section 10.3**, identify the word or concept being defined. Choose from the following words (some may be used more than once or not at all). *All answers are found in the Pre-Class Prep Answer Section.*

vertices	asymptotes	central rectangle
foci	hyperbola	center
$\dfrac{y^2}{a^2} - \dfrac{x^2}{b^2} = 1$	$4x^2 - y^2 = 16$	$\dfrac{x^2}{a^2} - \dfrac{y^2}{b^2} = 1$
$\dfrac{(x-h)^2}{a^2} - \dfrac{(y-k)^2}{b^2} = 1$	$\dfrac{(y-k)^2}{a^2} - \dfrac{(x-h)^2}{b^2} = 1$	discovery of the atomic structure of matter

1 Define a Hyperbola

1. The set of all points in a plane for which the difference of the distances from two fixed points is

   a constant:_____

2. Two fixed points from which any point on the hyperbola is a constant difference $d_1 - d_2$:_____

3. Midway between the foci:_____

Chapter 10 Conic Sections; More Graphing

2 Graph Hyperbolas Centered at the Origin

4. Points where the hyperbola crosses the x-axis: _____

5. Rectangle whose sides pass horizontally through $\pm b$ on the y-axis and vertically through $\pm a$ on

   the x-axis: _____

6. Pair of intersecting straight lines that the branches of the hyperbola get closer to as the

   hyperbola gets farther away from the origin: _____

7. Standard form of the equation of a horizontal hyperbola centered at the origin: _____

8. Standard form of the equation of a vertical hyperbola centered at the origin: _____

3 Graph Hyperbolas Centered at (h, k)

9. Standard equation of a hyperbola centered at (h, k) that opens left and right: _____

10. Standard equation of a hyperbola centered at (h, k) that opens up and down: _____

4 Graph Equations of the Form xy = k

11. The graph of the equation $xy = k$, where $k \neq 0$: _____

5 Solve Application Problems Involving Hyperbolas

12. Experiment of shooting high-energy alpha particles toward a thin sheet of gold: _____

13. The hyperbolic path of an alpha particle that is repelled by the nucleus at the origin:_____

Identify Conic Sections by their Equations

14. Graph of $4x^2 - 9y^2 = 144$: _____

15. Graph of $x^2 + y^2 = 16$: _____

16. Graph of $4x^2 + 9y^2 = 144$: _____

17. Graph of $x = y^2 + y - 16$: _____

Getting Ready for Class

Briefly look through the section again. Answer the following by writing the concept or just the page number from the text.

Identify concepts/procedures that you feel confident about:

Identify concepts/procedures that look confusing or challenging:

Be sure to ask your instructor further questions if you are still having difficulty with a concept.

10.3 The Hyperbola: In-Class Notes

Terms, Definitions, and Main Ideas	Examples and Notes

Use notebook paper for additional notes

10.3 The Hyperbola: After Class

Important to Know

What is your homework assignment? Be sure to note it in your weekly schedule.

Section 10.3 Homework: _____ **Due**: _____

Getting to Work!

Complete your homework assignment. If you are unable to do a problem, write down the problem number and a question to help you remember what you would like to ask your instructor, your tutor, or another student.

Problem Number	Question? Where in the problem did you start to have difficulty or confusion?	Answered?

Often, you will have more questions than there is space provided here. If so, write them on notebook paper and be sure to talk to your instructor. You might ask in class or privately with the instructor.

Do You Really Know It?

Can you put into words the concepts that you learned in this section? Answer the below question from the *Writing* section in the *Study Set* in your text. Explain as if you were explaining to someone who has never taken this class before. Use notebook paper if you need more room.

Explain why the graph of $\dfrac{x^2}{a^2} - \dfrac{y^2}{b^2} = 1$ *has no y-intercept.*

You Write the Test!

If you were writing the test for this section, what would you want a student to know? Write two test questions that you think might come from this material. Write questions of various difficulty; these questions can be original or chosen from the homework. Be sure to supply the answer also!

Write these questions at the end of this chapter under the section titled My Practice Chapter 10 Test and the answer to each question under the section titled My Practice Chapter 10 Test Answers.

Reflect on the Section

Look back at the *Pre-Class Prep* section. Did the lecture explain topics that you thought were going to be challenging or confusing? _____

- Are there topics that you still have questions on from the reading or the lecture? If so, complete the following:
    I don't understand…

- Speak to your instructor in class or during office hours about these concerns.

Reflect on Your Math Attitude

Consider the question: *Did I ever seek extra help from a tutor or from my instructor?* Describe your experiences with seeking help from your instructor or tutor. Were they positive, negative, neutral? If you did not seek help, explain why.

Identify what you would like to do in your next mathematics course to ensure that you receive adequate help.

Plan now to identify resources that will help you to succeed in your next mathematics course.

Are You Ready?

Complete the following problems. These review some basic skills that are needed when solving nonlinear systems of equations. *All answers are found in the Pre-Class Prep Answer Section.*

1. Graph: $2x - 3y = 6$

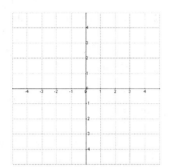

2. Solve: $9y^2 + 4y - 5 = 0$

3. Solve: $\begin{cases} 3x + 2y = 36 \\ 4x - y = 4 \end{cases}$

4. Solve: a. $x^2 = \dfrac{5}{9}$ b. $y^2 = 18$

Reading Time!

While **reading Section 10.4**, identify the statements as True or False. *All answers are found in the Pre-Class Prep Answer Section.*

1 Solve Systems by Graphing

_____ 1. An ordered pair of real numbers that satisfies all of the equations in the system is a solution of a nonlinear system of equations.

_____ 2. A secant line intersects a circle at one point and a tangent line intersects a circle at two points.

2 Solve Systems by Substitution

_____ 3. It is easy to determine the coordinates of the intersection points when solving a system by graphing.

_____ 4. A precise algebraic method called the substitution method can be used to solve certain systems involving nonlinear equations.

3 Solve Systems by Elimination (Addition)

_____ 5. The elimination or addition method is used most often when the equations of a nonlinear system are both first-degree equations.

_____ 6. With a system of nonlinear equations, it is possible to have four solutions.

✓ Getting Ready for Class

Briefly look through the section again. Answer the following by writing the concept or just the page number from the text.

Identify concepts/procedures that you feel confident about:

Identify concepts/procedures that look confusing or challenging:

Be sure to ask your instructor further questions if you are still having difficulty with a concept.

Terms, Definitions, and Main Ideas

Examples and Notes

Use notebook paper for additional notes

Chapter 10 Conic Sections; More Graphing

10.4 Solving Nonlinear Systems of Equations: After Class

What is your homework assignment? Be sure to note it in your weekly schedule.

Section 10.4 Homework: _____ **Due**: _____

Getting to Work!

Complete your homework assignment. If you are unable to do a problem, write down the problem number and a question to help you remember what you would like to ask your instructor, your tutor, or another student.

Problem Number	Question? Where in the problem did you start to have difficulty or confusion?	Answered?

Often, you will have more questions than there is space provided here. If so, write them on notebook paper and be sure to talk to your instructor. You might ask in class or privately with the instructor.

Do You Really Know It?

Can you put into words the concepts that you learned in this section? Answer the below question from the *Writing* section in the *Study Set* in your text. Explain as if you were explaining to someone who has never taken this class before. Use notebook paper if you need more room.

A. Describe the benefits of the graphical method for solving a system of nonlinear equations.

B. Describe the drawbacks of the graphical method.

You Write the Test!

If you were writing the test for this section, what would you want a student to know? Write two test questions that you think might come from this material. Write questions of various difficulty; these questions can be original or chosen from the homework. Be sure to supply the answer also!

Write these questions at the end of this chapter under the section titled *My Practice Chapter 10 Test* and the answer to each question under the section titled *My Practice Chapter 10 Test Answers*.

Reflect on the Section

Look back at the *Pre-Class Prep* section. Did the lecture explain topics that you thought were going to be challenging or confusing? _____

- Are there topics that you still have questions on from the reading or the lecture? If so, complete the following:
    I don't understand…

- Speak to your instructor in class or during office hours about these concerns.

Reflect on Your Math Attitude

Consider the question: *If I had it to do over, would I do anything differently?*
What would you do differently or what did you do to make this course a positive experience?

Plan now to implement steps to ensure a positive experience in your next mathematics course.

Chapter 10 Activities

 Your instructor may assign these activities to you to complete in class, or you may complete them on your own to solidify your understanding of chapter topics. The activities begin on the next page.

❖ **Student Activity:** *Constructing an Ellipse*

        Construct an ellipse using tacks, string and pen.

❖ **Student Activity:** *Name that Conic!*

        Match equations in the squares of a grid with their conic classifications.

❖ **Student Activity:** *Nonlinear System Detective*

        Determine the equations of a system of two equations where one is nonlinear, then solve the system.

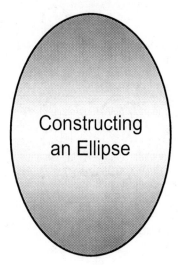

Constructing an Ellipse

Supplies needed for one construction:
- Bulletin Board
- Tacks (4)
- Unlined paper (the bigger the better)
- String (length less than the length of the paper)
- Felt tip pen (ballpoint will puncture paper on bulletin board)
- Straightedge

Definition of an Ellipse: An **ellipse** is the *set of all points* in a plane for which the *sum of the distances from two fixed points* is a constant. The fixed points are called the **foci** of the ellipse. The **center** is at the midpoint of the two foci. The line segment that forms the longest distance across the ellipse is called the **major axis**. The line segment that forms the smallest distance across the ellipse is called the **minor axis**. The endpoints of the major axis are called **vertices**.

Construct an Ellipse:
1. Use two tacks to affix your paper to the bulletin board.
2. Use the other two tacks to affix the ends of the string to the piece of paper. Place these to the left and right of the center of the paper (as shown). Leave some slack In the string (do not pull it taut).
3. Use the felt-tip pen to pull the string taut on the upper half of the paper (as shown). Move the pen to trace a curve, keeping the string taut.
4. After drawing half of the ellipse you will have to pick up the pencil and repeat the previous step on the lower half of the paper.
5. Remove your drawing from the bulletin board.

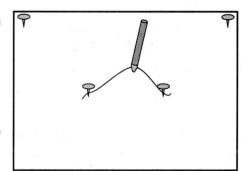

Add labels:

- The pen traced the *set of all points* in a plane for which the sum of the distances from two fixed points is a constant (the length of the string remained constant). Label your diagram with **Ellipse**.

- The points where the string was held down by a tack were the foci of the ellipse. Label each of these points with the word **focus**.

- Mark the point that is exactly halfway between the two foci and label it the **center**.

- Draw a line segment through the foci and center that extends to the ellipse curve. Label this line segment the **major axis**.

- Draw a line segment through the center that is perpendicular to the major axis that extends to the ellipse curve. Label this line segment the **minor axis**.

- Label each point at the end of the major axis with **vertex**.

Chapter 10 Conic Sections; More Graphing

Student Activity
Name that Conic!

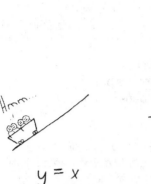

$y = x$

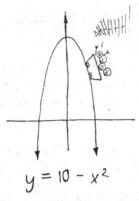

$y = 10 - x^2$

... Conics are the BEST!!!!!!

Match-up: Match each of the equations in the squares of the grid below with the conic classification. You may have to rearrange or simplify equations to determine their classification.

A Circle
B Ellipse
C Parabola opening up or down
D Parabola opening right or left

E Hyperbola opening right and left
F Hyperbola opening up and down
G Something else

$\dfrac{x^2}{25} - \dfrac{y^2}{9} = 1$	$x = y^2 + 1$	$x^2 + 4y^2 = 4$
$x^2 + 3(y-3)^2 = 9$	$x^2 + (y-3)^2 = 1$	$x^2 + (y-3)^2 = 9$
$y - (x-4)^2 = 5$	$y = \dfrac{3}{x}$	$9(y-4)^2 - 4(x+1)^2 = 36$
$x + (y+9)^2 = 3$	$x^2 + y^2 = 1$	$x^2 + 4x - y^2 + 2y = 6$
$x^2 + 5x + 3 - (x^2 + 2x) = y$	$25x^2 + 100x - 4y^2 = 100$	$\dfrac{x^2}{4} + \dfrac{y^2}{4} = 1$

Student Activity
Nonlinear System Detective

Instructions: Each graph shows a system of two equations where one of the equations is nonlinear. Write the equations then solve the system algebraically.

1.

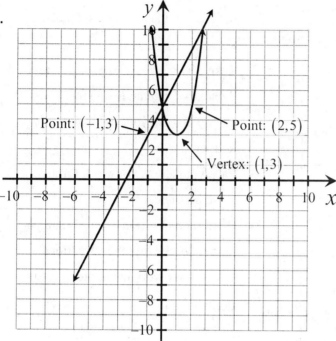

Equations:

Solve:

2.

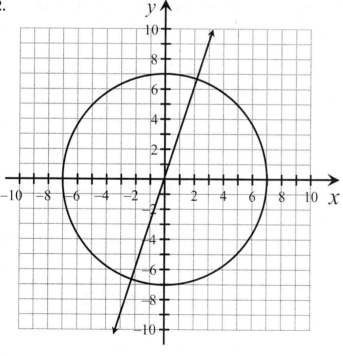

Equations:

Solve:

Chapter 10 Conic Sections; More Graphing

Chapter 10 Test Skills Assessment

Pre-Test Preparation Work:

1. Re-read the objectives from each section.

2. Review the *Reading Time!* activity for each section.

3. Go over all your classroom notes, if something in your notes doesn't make sense to you, make a note and ask your teacher or a classmate.

4. Make additional notations to your work if your teacher states specific concepts to study in preparation for the chapter test.

5. Attend any study sessions held by your teacher or teaching assistant.

6. Practice additional problems.

7. Go over any missed problems in your homework sets.

8. Talk out concepts with your peers in small group study sessions.

List other preparations that you have found beneficial in preparing for a math test.

Additional Practice Suggestions

1. Use your book's review problems at the back of the chapter as a practice test.

2. Take *My Practice Chapter Test* and the text's *Chapter Test*. Time yourself and do not use your notebook or textbook.

3. Pace yourself as you work through these problems.

4. Read each question carefully, playing close attention to the instructions.

5. Check your work using the answers provided.

6. Rework any missed problems. Do not just "look them over" but actually rework the problem without looking at text or notes.

My Practice Chapter 10 Test

For each section, you had the opportunity to create two test questions under the section *You Write the Test!* Write each of those questions here. Include your answers under the heading *My Practice Chapter 10 Test Answers*. **Take the test without notes or your textbook.** If you do not get a question correct, review the text and/or your notes then take the test again. For further review, do the *Chapter 10 Test* in the text.

Section 10.1

1.

2.

Section 10.2

3.

4.

Section 10.3

5.

6.

Section 10.4

7.

8.

Chapter 10 Conic Sections; More Graphing

Section 10.1

1.

2.

Section 10.2

3.

4.

Section 10.3

5.

6.

Section 10.4

7.

8.

Chapter 11 Miscellaneous Topics

Read the *Study Skills Workshop* found at the beginning of Chapter 11 in your textbook. **Complete** the activities below for this chapter's *Study Skills Workshop*.

Exploring Careers

Your career path will determine the math course(s) that you will need to take after Intermediate Algebra.

How Do You Decide?

See the advice of a counselor, visit your school's career center, search the Internet, or read books that will help you discover your interests and possible related careers.

✓ *Do you have a career goal in mind? If so, what is it?*

_____

_____

Take at least two personality tests and two career-choice tests. A list of tests offered online can be found at **www.cengage.com/math/tussy**.

✓ *Write down the names and websites of the tests that you would like to take.*

Personality Tests	Website

Career-choice Tests	Website

Once You've Decided

Talk to your counselor and consult the appropriate college catalogs to develop a long-term plan that will put you on the correct educational path.

✓ *Write down the classes suggested by your counselor. Use more paper if necessary.*

11.1 The Binomial Theorem: Pre-Class Prep

✓ Are You Ready?

Complete the following problems. These review some basic skills that are needed when working with the binomial theorem. *All answers are found in the Pre-Class Prep Answer Section.*

1. Determine the coefficient of each term: $x^4 + 4x^3 + 6x^3 + 4x + 1$

2. Find the product: $(a+b)^2$

3. Evaluate: $5 \cdot 4 \cdot 3 \cdot 2 \cdot 1$

4. Simplify: $\dfrac{10 \cdot 9 \cdot 8}{2 \cdot 3 \cdot 4}$

✓ Reading Time!

While **reading Section 11.1**, fill in the blanks choosing from the following words (some may be used more than once or not at all). *All answers are found in the Pre-Class Prep Answer Section.*

sum	1	above	binomial
$n!$	decrease	$(a+b)^n$	increase
0	Pascal's	expand	product
$\dfrac{n!}{r!(n-r)!}a^{n-r}b^r$	$n(n-1)(n-2)(n-3)\cdots\cdot 3 \cdot 2 \cdot 1$		

1 Use Pascal's Triangle to Expand Binomials

1. To _____ a binomial of the form $(a+b)^n$, where n is a nonnegative integer, means to write it as a _____ of terms.

2. _____ triangle is an array where each entry between the 1's is the _____ of the closest pair of numbers in the line immediately _____ it.

2 Use Factorial Notation

3. Factorial notation is _____ where $n!$ is the _____ of consecutively decreasing natural numbers from n to 1. Specifically, $n! =$_____.

4. Zero factorial is defined as $0! =$_____.

5. For any natural number n, $n(n-1)! =$_____ .

3 Use the Binomial Theorem to Expand Binomials

6. The Binomial Theorem states: For any positive integer n,

$$\underline{\hspace{1.5cm}} = a^n + \frac{n!}{1!(n-1)!}a^{n-1}b + \frac{n!}{2!(n-2)!}a^{n-2}b^2 + \frac{n!}{3!(n-3)!}a^{n-3}b^3 + \cdots + \frac{n!}{r!(n-r)!}a^{n-r}b^r + \cdots + b^n$$

7. In the _____ theorem, the exponents on the variables follow the familiar pattern:

- The _____ of the exponents on a and b in each term is n.

- The exponents on a _____ by 1 in each successive term.

- The exponents on b _____ by 1 in each successive term.

4 Find a Specific Term of a Binomial Expansion

8. The $(r+1)^{st}$ term of the expansion of $(a+b)^n$ is _____.

Getting Ready for Class

Briefly look through the section again. Answer the following by writing the concept or just the page number from the text.

Identify concepts/procedures that you feel confident about:

Identify concepts/procedures that look confusing or challenging:

Be sure to ask your instructor further questions if you are still having difficulty with a concept.

Chapter 11 Miscellaneous Topics

11.1 The Binomial Theorem: In-Class Notes

Terms, Definitions, and Main Ideas	Examples and Notes

Use notebook paper for additional notes

11.1 The Binomial Theorem: After Class

Important to Know

What is your homework assignment? Be sure to note it in your weekly schedule.

Section 11.1 Homework: _____ Due: _____

Getting to Work!

Complete your homework assignment. If you are unable to do a problem, write down the problem number and a question to help you remember what you would like to ask your instructor, your tutor, or another student.

Problem Number	Question? Where in the problem did you start to have difficulty or confusion?	Answered?

Often, you will have more questions than there is space provided here. If so, write them on notebook paper and be sure to talk to your instructor. You might ask in class or privately with the instructor.

Do You Really Know It?

Can you put into words the concepts that you learned in this section? Answer the below question from the *Writing* section in the *Study Set* in your text. Explain as if you were explaining to someone who has never taken this class before. Use notebook paper if you need more room.

Explain why the signs alternate in the expansion of $(x - y)^9$.

You Write the Test!

If you were writing the test for this section, what would you want a student to know? Write two test questions that you think might come from this material. Write questions of various difficulty; these questions can be original or chosen from the homework. Be sure to supply the answer also!

Write these questions at the end of this chapter under the section titled *My Practice Chapter 11 Test* and the answer to each question under the section titled *My Practice Chapter 11 Test Answers*.

 Reflect on the Section

Look back at the *Pre-Class Prep* section. Did the lecture explain topics that you thought were going to be challenging or confusing? _____

- Are there topics that you still have questions on from the reading or the lecture? If so, complete the following:
    I don't understand…

- Speak to your instructor in class or during office hours about these concerns.

 Reflect on Your Math Attitude

Which statement best completes this statement: The number of math courses required…

_____ will have no effect on my career choice.

_____ will affect my career choice towards careers that require little to no math.

Have you decided on a career path? How might your attitude towards mathematics affect your career choice?

Try to develop a positive attitude towards mathematics so that you are not limited in your career options.

11.2 Arithmetic Sequences and Series: Pre-Class Prep

Are You Ready?

Complete the following problems. These review some basic skills that are needed when working with arithmetic sequences and series. *All answers are found in the Pre-Class Prep Answer Section.*

1. Let $f(x) = 6x - 3$. Find $f(4)$.

2. Evaluate: a. $(-1)^4$ b. $(-1)^5$

3. Find the difference: $11 - 3$

4. Evaluate: $12 + (6-1)8$

Reading Time!
(on the next page)

Getting Ready for Class

Briefly look through the section again. Answer the following by writing the concept or just the page number from the text.

Identify concepts/procedures that you feel confident about:

Identify concepts/procedures that look confusing or challenging:

Be sure to ask your instructor further questions if you are still having difficulty with a concept.

Chapter 11 Miscellaneous Topics

Reading Time!

While **reading Section 11.2**, match the word or concept to its definition or description. Not all choices are used. *All answers are found in the Pre-Class Prep Answer Section.*

1 Find Terms of a Sequence Given the General Term

_____ 1. Finite sequence

_____ 2. Term

_____ 3. Infinite sequence

_____ 4. General term

2 Find Terms of an Arithmetic Sequence by Identifying the First Term and the Common Difference

_____ 5. Arithmetic sequence

_____ 6. Common difference

_____ 7. General term formula

3 Find Arithmetic Means

_____ 8. Arithmetic means

4 Find the Sum of the First n Terms of an Arithmetic Sequence

_____ 9. Sum of the first n terms of an arithmetic sequence

5 Solve Application Problems Involving Arithmetic Sequences

_____ 10. Used to find the number of band members

6 Use Summation Notation

_____ 11. Arithmetic series

_____ 12. Summation notation

_____ 13. Index of the summation

A. sequence in which each term is found by adding the same number to the previous term

B. $S_n = \dfrac{n(a_1 + a_n)}{2}$

C. each number in a sequence

D. $a_n = a_1 + (n-1)d$

E. sum of the terms of an arithmetic sequence

F. function whose domain is the set of natural numbers $\{1,2,3,4,\ldots,n\}$ for some natural number n (finite number of terms)

G. k in the expression $\displaystyle\sum_{k=1}^{4} 3k$

H. function whose domain is the set of natural numbers $\{1,2,3,4,\ldots\}$ (infinite number of terms)

I. d, the difference between any two consecutive terms in an arithmetic sequence

J. a_n, describes all the terms of the sequence

K. arithmetic sequence

L. notation to write a series, involves the Greek letter Σ

M. numbers inserted between two numbers a and b to form an arithmetic sequence.

Terms, Definitions, and Main Ideas	Examples and Notes

Use notebook paper for additional notes.

11.2 Arithmetic Sequences and Series: After Class

Important to Know

What is your homework assignment? Be sure to note it in your weekly schedule.

Section 11.2 Homework: _____ **Due**: _____

Getting to Work!

Complete your homework assignment. If you are unable to do a problem, write down the problem number and a question to help you remember what you would like to ask your instructor, your tutor, or another student.

Problem Number	Question? Where in the problem did you start to have difficulty or confusion?	Answered?

Often, you will have more questions than there is space provided here. If so, write them on notebook paper and be sure to talk to your instructor. You might ask in class or privately with the instructor.

Do You Really Know It?

Can you put into words the concepts that you learned in this section? Answer the below question from the *Writing* section in the *Study Set* in your text. Explain as if you were explaining to someone who has never taken this class before. Use notebook paper if you need more room.

What is the difference between a sequence and a series?

You Write the Test!

If you were writing the test for this section, what would you want a student to know? Write two test questions that you think might come from this material. Write questions of various difficulty; these questions can be original or chosen from the homework. Be sure to supply the answer also!

Write these questions at the end of this chapter under the section titled *My Practice Chapter 11 Test* and the answer to each question under the section titled *My Practice Chapter 11 Test Answers*.

Reflect on the Section

Look back at the *Pre-Class Prep* section. Did the lecture explain topics that you thought were going to be challenging or confusing? _____

- Are there topics that you still have questions on from the reading or the lecture? If so, complete the following:
  I don't understand…

- Speak to your instructor in class or during office hours about these concerns.

Reflect on Your Math Attitude

In the *Study Skills Workshop* for this chapter, you were encouraged to take at least two personality tests and two career-choice tests. Describe what you learned with respect to your attitude towards mathematics from each test.

Personality Tests:

Career-choice Tests:

Let the results of these tests give you direction in deciding your career path.

11.3 Geometric Sequences and Series: Pre-Class Prep

 Are You Ready?

Complete the following problems. These review some basic concepts that are needed when working with geometric sequences and series. *All answers are found in the Pre-Class Prep Answer Section.*

1. Let $f(x) = 6(2^x)$. Find $f(3)$.

2. Evaluate: a. $(5)^4$ b. $\left(\dfrac{1}{3}\right)^3$

3. Simplify: $\dfrac{6}{24}$

4. Solve: $r^2 = 144$

 Reading Time!

While **reading Section 11.3**, identify the word or concept being defined. Choose from the following words (some may be used more than once or not at all). *All answers are found in the Pre-Class Prep Answer Section.*

use a geometric sequence	common ratio	partial sum
first term	geometric means	geometric series
$a_n = a_1 r^{n-1}$	$r = \dfrac{a_{n+1}}{a_n}$	$S = \dfrac{a_1}{1-r}$

find the sum of the terms of the infinite geometric sequence $S_n = \dfrac{a_1 - a_1 r^n}{1-r}$, where $r \neq 1$

1 Find Terms of a Geometric Sequence by Identifying the First Term and the Common Ratio

1. A sequence of the form $a_1, a_1 r, a_1 r^2, a_1 r^3, \dots, a_1 r^{n-1}, \dots$:_____

2. General form formula for the nth term of a geometric sequence: _____

3. In the geometric sequence $a_1, a_1 r, a_1 r^2, a_1 r^3, \dots, a_1 r^{n-1}, \dots$, a_1 is the:_____

4. The quotient obtained when any term is divided by the previous term:_____

2 Find Geometric Means

5. The numbers inserted between two numbers *a* and *b* to form a geometric sequence: _____

3 Find the Sum of the First n Terms of a Geometric Sequence

6. The sum of the first *n* terms of a geometric sequence: _____

4 Define and Use Infinite Geometric Series

7. The sum of the terms of a geometric sequence: _____

8. The sum of the terms of an infinite geometric sequence: _____

9. A portion of the sum of a geometric series: _____

10. Sum of the terms of an infinite geometric series formula: _____

5 Solve Application Problems Involving Geometric Sequences

11. To find the amount of money left in the inheritance fund after 20 years of payments: _____

12. To find the total distance the bearing will travel: _____

✓ Getting Ready for Class

Briefly look through the section again. Answer the following by writing the concept or just the page number from the text.

Identify concepts/procedures that you feel confident about:

Identify concepts/procedures that look confusing or challenging:

Be sure to ask your instructor further questions if you are still having difficulty with a concept.

11.3 Geometric Sequences and Series: In-Class Notes

Terms, Definitions, and Main Ideas	Examples and Notes

Use notebook paper for additional notes

11.3 Geometric Sequences and Series: After Class

Important to Know

What is your homework assignment? Be sure to note it in your weekly schedule.

Section 11.3 Homework: _____ **Due:** _____

Getting to Work!

Complete your homework assignment. If you are unable to do a problem, write down the problem number and a question to help you remember what you would like to ask your instructor, your tutor, or another student.

Problem Number	Question? Where in the problem did you start to have difficulty or confusion?	Answered?

Often, you will have more questions than there is space provided here. If so, write them on notebook paper and be sure to talk to your instructor. You might ask in class or privately with the instructor.

Do You Really Know It?

Can you put into words the concepts that you learned in this section? Answer the below question from the *Writing* section in the *Study Set* in your text. Explain as if you were explaining to someone who has never taken this class before. Use notebook paper if you need more room.

Why must the absolute value of the common ratio be less than 1 before an infinite geometric sequence can have a sum?

You Write the Test!

If you were writing the test for this section, what would you want a student to know? Write two test questions that you think might come from this material. Write questions of various difficulty; these questions can be original or chosen from the homework. Be sure to supply the answer also!

Write these questions at the end of this chapter under the section titled *My Practice Chapter 11 Test* and the answer to each question under the section titled *My Practice Chapter 11 Test Answers*.

Reflect on the Section

Look back at the *Pre-Class Prep* section. Did the lecture explain topics that you thought were going to be challenging or confusing? _____

- Are there topics that you still have questions on from the reading or the lecture? If so, complete the following:
  I don't understand…

- Speak to your instructor in class or during office hours about these concerns.

Reflect on Your Math Attitude

The *Study Skills Workshop* for this chapter encourages you to talk to your counselor and develop a long-term plan for your chosen career choice. Does your future plan involve additional mathematics courses?

Look back at your overall attitude in this course. Do you feel that your attitude helped or hindered your performance? If you need to take more mathematics courses, identify what you plan to do in order to keep a positive attitude or to improve your attitude towards math.

Plan now to identify resources that will help you to succeed in your chosen career path.

Chapter 11 Activities

Your instructor may assign these activities to you to complete in class, or you may complete them on your own to solidify your understanding of chapter topics. The activities begin on the next page.

❖ **Student Activity:** *A Sequence of Tables*

Fill in the missing information to demonstrate your knowledge of arithmetic sequences.

❖ **Student Activity:** *Infinitely Finite*

Investigate how it is possible that an infinite series can be bounded by a finite number.

Chapter 11 Miscellaneous Topics

Student Activity:
A Sequence of Tables

...84, 88, 92, 96, 100, 104, 108, ...

I think he's counting to n=100 with the sequence $a_n = 4n$.

Billy loved playing hide and seek ... especially when he got to be it.

Directions: Fill in the missing information in the table below to demonstrate your knowledge of arithmetic sequences.

Level 1: Sequences in General				
Sequence of numbers	**# of terms**	**State this term**	**General Term**	**Find this term**
$-1, 4, 9, 14...$		$a_3 =$	$a_n = 5n - 6$	$a_{10} =$
$0, 5, 10, 15, 20, 25, 30$		$a_4 =$	$a_n =$	$a_6 =$
	5	$a_2 =$	$a_n = \dfrac{(-1)^n}{5^n}$	$a_5 =$
	Infinite	$a_5 =$	$a_n = 2^n - 1$	$a_8 =$

Level 2: Infinite Arithmetic Sequences				
Sequence of numbers	$d =$	$a_1 =$	$a_n =$	$a_{20} =$
$6, 12, 18, 24,...$				
			$a_n = -2 + 5n$	
	5	8		
	3			$a_{20} = 66$

Level 3: Arithmetic Series					
Series of numbers	$d =$	$a_1 =$	$a_n =$	**Find this term**	**Find this sum**
$4 + 12 + 20 + ...$				$a_{20} =$	$S_{20} =$
$50 + 55 + 60 + ...$				$a_{12} =$	$S_{12} =$
	3	-2		$a_{10} =$	$S_{10} =$
			$a_n = 4 + 3n$	$a_{100} =$	$S_{100} =$

Cengage Student Workbook Activities, M. Andersen

Student Activity
Infinitely Finite

OK, a little to the right...
Just a little more in that direction ...
A little more ...

It can be very difficult to grasp the idea that an infinite series can be bounded by a finite number. Here are two examples that might help you to clarify that, at the very least, it is possible.

After scratching for what seemed like an eternity, he suddenly understood "infinitely finite."

Example 1: If each large square has an area of 1 square unit, what is the size of each of the smaller squares or rectangles below?

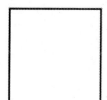 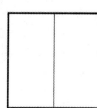 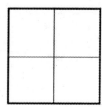 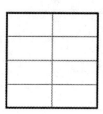 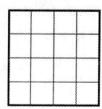

a. Area: _1_   Area: ____   Area: ____   Area: ____   Area: ____

Now we illustrate the series: $\frac{1}{2}+\frac{1}{4}+\frac{1}{8}+...$ by shading in the appropriate pieces of the square as we add each term of the series.

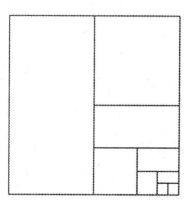

b. Based on this shading, what is the sum of the infinite series $\frac{1}{2}+\frac{1}{4}+\frac{1}{8}+...$? _____

c. Now show how to find this sum using $S=\frac{a}{1-r}$.

Cengage Student Workbook Activities, M. Andersen

Chapter 11 Miscellaneous Topics

Example 2: Now we look at the series $\dfrac{3}{10} + \dfrac{3}{100} + \dfrac{3}{1,000} + \dfrac{3}{10,000} + \ldots$

a. $a_1 = $ _____ and $r = $ _____ .

b. Find the sum of the infinite series using $S = \dfrac{a_1}{1-r}$.

c. Now write the same series using decimals instead of fractions:

_____ + _____ + _____ + _____ + _____ + ...

d. You should be able to clearly see the sum of the series now without any sort of calculation. Write the sum of the series as a decimal: _____

Example 2: Now we look at the series $\dfrac{7}{10} + \dfrac{7}{100} + \dfrac{7}{1,000} + \dfrac{7}{10,000} + \ldots$

a. $a_1 = $ _____ and $r = $ _____ .

b. Find the sum of the infinite series using $S = \dfrac{a_1}{1-r}$.

c. Now write the same series using decimals instead of fractions:

_____ + _____ + _____ + _____ + _____ + ...

d. You should be able to clearly see the sum of the series now without any sort of calculation. Write the sum of the series as a decimal: _____

Chapter 11 Test Skills Assessment

Pre-test Preparation Work:

1. Re-read the objectives from each section.

2. Review the *Reading Time!* activity for each section.

3. Go over all your classroom notes, if something in your notes doesn't make sense to you, make a note and ask your teacher or a classmate.

4. Make additional notations to your work if your teacher states specific concepts to study in preparation for the chapter test.

5. Attend any study sessions held by your teacher or teaching assistant.

6. Practice additional problems.

7. Go over any missed problems in your homework sets.

8. Talk out concepts with your peers in small group study sessions.

List other preparations that you have found beneficial in preparing for a math test.

Additional Practice Suggestions

1. Use your book's review problems at the back of the chapter as a practice test.

2. Take *My Practice Chapter Test* and the text's *Chapter Test*. Time yourself and do not use your notebook or textbook.

3. Pace yourself as you work through these problems.

4. Read each question carefully, playing close attention to the instructions.

5. Check your work using the answers provided.

6. Rework any missed problems. Do not just "look them over" but actually rework the problem without looking at text or notes.

My Practice Chapter 11 Test

For each section, you had the opportunity to create two test questions under the section *You Write the Test!* Write each of those questions here. Include your answers under the heading *My Practice Chapter 11 Test Answers*. **Take the test without notes or your textbook.** If you do not get a question correct, review the text and/or your notes then take the test again. For further review, do the *Chapter 11 Test* in the text.

Section 11.1

1.

2.

Section 11.2

3.

4.

Section 11.3

5.

6.

My Practice Chapter 11 Test Answers

Section 11.1

1.

2.

Section 11.2

3.

4.

Section 11.3

5.

6.

Chapter 11 Miscellaneous Topics

Pre-Class Prep Answer Section

Answers to **Are You Ready?** questions are also located in the answer section in the back of the textbook or at www.webassign.net/brookscole.

Section 1.1
Are You Ready?
1. a. addition b. subtraction c. multiplication d. division 2. 20 3. 32 4. 11
Reading Time!
1. product, sum 2. dividing, subtracting 3. words
4. variables, constant 5. expressions 6. mathematical
7. equation 8. formula 9. two-column table

Section 1.2
Are You Ready?
1. 34 2. –27 3. a decimal (3.6 GPA)
4. a fraction ($\frac{3}{4}$ cup of flour)
Reading Time!
1. True   2. True   3. False   4. False   5. False
6. True   7. True   8. True   9. True   10. False
11. True 12. True   13. True 14. True 15. False
16. False 17. True   18. False 19. True 20. True
21. True 22. True   23. False 24. True 25. False

Section 1.3
Are You Ready?
1. 4, 6; –6 has the larger absolute value 2. 12.17 3. 11
4. $\frac{5}{18}$ 5. $\frac{4}{9}$ 6. 8.5
Reading Time!
1. O 2. K 3. N 4. H 5. F 6. M 7. A 8. J 9. B
10. D 11. G 12. C 13. I 14. E

Section 1.4
Are You Ready?
1. The factors are in a different order. 2. The position of the parentheses is different. 3. 42, 42; same result
4. Different variable factors (x and y); the same coefficient (8)
Reading Time!
1. term 2. constant term 3. coefficient
4. Commutative Properties 5. equivalent expressions
6. additive identity property 7. multiplicative identity property 8. multiplication property of 0 9. additive inverse property 10. multiplicative inverse property
11. division of a number by itself 12. division by 1
13. division of 0 14. division by 0 15. division of 0 by 0 16. Simplify 17. Commutative Properties of Multiplication, Associative Properties of Multiplication 18. distributive property 19. $ab - ac$
20. extended distributive property 21. like terms
22. unlike terms 23. combining like terms

Section 1.5
Are You Ready?
1. –6 2. 10a 3. –11m+8 4. x 5. 18n 6. 27
Reading Time!
1. solution set 2. variable 3. satisfies 4. solution
5. equation 6. right side 7. linear equation in one variable 8. equivalent equations 9. Addition Property of Equality 10. Subtraction Property of Equality 11. Multiplication Property of Equality
12. Division Property of Equality 13. isolated
14. clear equation of fractions and decimals
15. simplify each side of the equation 16. isolate the variable term on one side 17. isolate the variable
18. check the result 19. multiply by LCD
20. multiply by a power of 10 21. Identity
22. Contradiction 23. empty set 24. $\mathbb{R}$

Section 1.6
Are You Ready?
1. volume 2. circumference 3. perimeter 4. area
Reading Time!
1. N 2. F 3. K 4. H 5. A 6. I 7. L 8. B 9. C
10. J 11. E 12. G 13. M 14. D

Section 1.7
Are You Ready?
1. 2x – 8 2. 93 3. $59.50 4. $P = 2l + 2w$ 5. 180°
6. 22
Reading Time!
1. analyze, facts 2. assign, unknown 3. equation, mathematical 4. solve 5. conclusion 6. check, original 7. number · value 8. table 9. right angle, straight angle 10. acute angle, obtuse angle 11. 90°, complement 12. supplementary, supplement
13. isosceles, vertex 14. right, equilateral 15. base angles 16. sum

Section 1.8
Are You Ready?
1. a. 0.278 b. 83.56% 2. $300 3. 220 mi 4. 0.6 L of alcohol 5. $33.30 6. $(60,000 – x)
Reading Time!
1. percent 2. percent sentence 3. markdown
4. regular price, markdown, percent 5. changed
6. original 7. $I = Prt$ 8. principal 9. uniform motion
10. $d = rt$ 11. dry mixture,
total value = amount · price 12. concentrations
13. amount of solution · strength of the solution

Pre-Class Prep Answer Section

Section 2.1
Are You Ready?

1.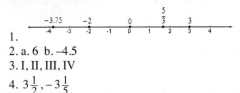
2. a. 6 b. –4.5
3. I, II, III, IV
4. $3\frac{1}{2}, -3\frac{1}{5}$

Reading Time!
1. quadrants 2. *x*-axis 3. coordinate plane
4. *y*-coordinate 5. *y*-axis 6. ordered pair 7. origin
8. coordinates 9. graphing/plotting 10. (–4, 6)
11. paired data 12. graph 13. step graph

14. Midpoint 15. Endpoints 16. $\left(\dfrac{x_1 + x_2}{2}, \dfrac{y_1 + y_2}{2}\right)$

Section 2.2
Are You Ready?
1. False 2. –3
3.
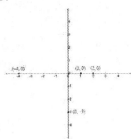

4. 0
5. –2
6.

Reading Time!
1. solution, ordered pair 2. two 3. satisfies
4. solution, select a number, corresponding
5. infinitely many, graph 6. all 7. three, plot, straight
line 8. straight line, linear 9. linear, $Ax + By = C$,
zero (0), standard 10. intercept method 11. (0,*b*),
y-axis, 0, *y* 12. (*a*, 0), *x*-axis, 0, *x* 13. *x*-axis
14. horizontal 15. vertical 16. is 17. real-life, linear
models, variables 18. straight-line depreciation,
declining, equation

Section 2.3
Are You Ready?
1. $\frac{3}{5}$ 2. undefined 3. $\frac{6}{7}$ 4. Parallel, Perpendicular

Reading Time!
1. D 2. A 3. R 4. Q 5. F 6. M 7. N 8. G 9. L 10. B
11. K 12. C 13. E 14. H 15. P 16. I 17. O

Section 2.4
Are You Ready?
1. a. 5*x*, –8 b. 5 2. $y = -\frac{4}{5}x + 4$ 3. The *y*-axis 4. $\frac{2}{3}$
5. $y = 2x - 23$ 6. $\frac{45}{8}$

Reading Time!
1. True 2. False 3. True 4. False 5. True 6. True
7. False 8. False 9. True 10. True 11. True 12. False
13. True 14. False 15. False 16. True 17. True
18. True 19. True 20. True

Section 2.5
Are You Ready?
1. (3, 5), (3, 0) 2. 7 3. 3, (0, –8) 4. $-\frac{3}{2}$

Reading Time!
1. I 2. N 3. C 4. D 5. M 6. A 7. G 8. J
9. F 10. O 11. B 12. H 13. L 14. E 15. K

Section 2.6
Are You Ready?
1. 3, 9 2. 1, 1 3. –8, 8 4. 4, 4
Reading Time!
1. value of *f(a)* 2. domain 3. range 4. nonlinear
functions 5. point-plotting method 6. squaring
function 7. parabola 8. cubing function
9. absolute value function 10. vertical translation
11. *k* units upward 12. *k* units downward
13. horizontal translation 14. *h* units to the left 15. *h*
units to the right 16. reflection 17. *x*-axis
18. vertical Line Test

Section 3.1
Are You Ready?
1. Not a solution
2.   3.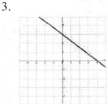

4. Parallel

Reading Time!
1. system of equations   2. solution of the system
3. solve   4. graphing, same   5. intersect, solution
6. no   7. check   8. consistent   9. independent
10. inconsistent   11. dependent   12. equivalent
systems   13. linear equations   14. graphing, x
15. graphing

Section 3.2
Are You Ready?
1. 1   2. $y = 5x + 4$ 3. –9   4. 3   5. 4
6. $-28x + 4y = -36$
Reading Time!
1. substitution method   2. substitution equation
3. step 3 of the substitution method   4. step 2 of the
substitution method   5. step 1 of the substitution
method   6. step 4 of the substitution method
7. elimination method   8. $Ax + By = C$ 9. Opposites
10. terms involving x (or y)   11. infinitely many
solutions   12. no solution   13. inconsistent
14. consistent and dependent   15. substitution method
16. elimination method   17. graphing method

Section 3.3
Are You Ready?
1. 1, –4, 5   2. 4   3. $7x + 5y - z = 9$ 4. (–5, 10)
Reading Time!
1. True   2. False   3. True   4. True   5. False   6. False
7. True   8. True   9. False   10. True   11. False   12. True
13. False   14. True   15. True

Section 3.4
Are You Ready?
1. a. 3, –7   b. 5, –1   2. 1   3. 1   4. $-4x + 12y = -10$
Reading Time!
1. matrix, rows   2. horizontal, vertical   3. element
4. order   5. augmented matrix   6. one equation,
coefficients, constants   7. augmented matrix, 1's,
main diagonal   8. reduced row-echelon form
9. interchanged, nonzero, adding   10. Gauss-Jordan
elimination   11. augmented matrix, reduced row-
echelon form, solution, original   12. Gauss-Jordan
elimination   13. inconsistent, no   14. dependent,
infinitely many

Section 3.5
Are You Ready?
1. 3 rows, 3 columns   2. 3, –1   3. –54   4. –123
Reading Time!
1. square matrix   2. determinant   3. $ad - bc$
4. expanding by minors

5. $a_1 \begin{vmatrix} b_2 & c_2 \\ b_3 & c_3 \end{vmatrix} - b_1 \begin{vmatrix} a_2 & c_2 \\ a_3 & c_3 \end{vmatrix} + c_1 \begin{vmatrix} a_2 & b_2 \\ a_3 & b_3 \end{vmatrix}$

6.

$+\ \ -\ \ +$
$-\ \ +\ \ -$
$+\ \ -\ \ +$

7. Cramer's rule   8. $x = \dfrac{D_x}{D}$, $y = \dfrac{D_y}{D}$

9. 0 (zero)   10. inconsistent   11. consistent

12. $x = \dfrac{D_x}{D}$, $y = \dfrac{D_y}{D}$, $z = \dfrac{D_z}{D}$ 13. D 14. D_y

Section 3.6
Are You Ready?
1. $w - 150$ 2. $1,040   3. 72 mi   4. 32 lb
Reading Time!
1. G   2. D   3. N   4. O   5. K   6. F   7. I   8. H   9. A
10. J   11. Q   12. L   13. P   14. B   15. E   16. R   17. C
18. M

Section 3.7
Are You Ready?
1. $75 + n$ 2. $150n   3. $5n$¢   4. $\left(-4, -4, 2 \right)$

Reading Time!
1. problem-solving strategy   2. solve, elimination,
matrices, Cramer's rule   3. table   4. revenue
5. unknowns, three, check   6. curve fitting

7. equation, $y = ax^2 + bx + c$, solve, three

Section 4.1
Are You Ready?
1. is greater than   2. False
3.

4. $0 < 6$
Reading Time!
1. H   2. L   3. B   4. G   5. F   6. D   7. Q   8. E
9. J   10. K   11. N   12. C   13. A   14. O   15. M
16. I

Section 4.2

Are You Ready?

1. 3 and 4   2. $[-7,\infty)$ 3. Yes

4

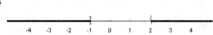

Reading Time!

1. intersection of set A and set B 3. union of set A and set B 3. Venn diagrams   4. compound inequality
5. all numbers that make both inequalities true
6. bounded interval   7. closed intervals 8. open intervals   9. half-open intervals   10. double inequalities 11. $c < x$ and $x < d$ 12. all numbers that make one or the other or both inequalities true
13. on the same number line   14. intersection
15. union 16. interval notation or set-builder notation

Section 4.3

Are You Ready?

1. a. 12 b. 7.5   2. a. True b. False

3. $(-\infty,-3] \cup [1,\infty)$ 4. $(-8,4)$

Reading Time!

1. True   2. True   3. False   4. True   5. True
6. False   7. False   8. True   9. True   10. True
11. False   12. False

Section 4.4

Are You Ready?

1. False 2. True

3.

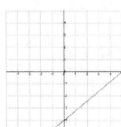

4. a. below b. above c. on

Reading Time!

1. B 2. I 3. M   4. F   5. G   6. K   7. D   8. H   9. C
10. E   11. A   12. J   13. L

Section 4.5

Are You Ready?

1. It is not a solution   2. vertical line

3.

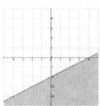

4. greater than or equal to

Reading Time!

1. systems of linear inequalities 2. Step 1. same; Step 2. intersection, solutions; Step 3. check, original
3. solution set 4. vertical 5. horizontal, above, below
6. $\leq$ 7. $\geq$

Section 5.1

Are You Ready?

1. a. 18 b. 216 2. a. 32 b. 10 3. x^6 4. a. y^2 b. 7^3
5. $\frac{1}{8}$ 6. $\cdot$ (multiplication dot)

Reading Time!

1. exponent 2. base 3. natural-number exponent
4. exponential expressions 5. 1 6. product rule for exponents 7. power rule for exponents 8. power of a product rule 9. power of a quotient rule 10. $x^0 = 1$
11. indeterminate form 12. negative exponents

13. x^n 14. x^n 15. $\frac{y^n}{x^m}$ 16. quotient rule for

exponents 17. rule for negative exponents and reciprocals
18. i. product rule ii. power rule iii. power of a product iv. quotient rule v. power of a quotient
vi. exponents of 0 and 1 vii. negative exponent
viii. negative exponents appearing in fractions

Section 5.2

Are You Ready?

1. 10,000 2. 9,630,000 3. $\frac{1}{1,000}$ 4. 0.0431

Reading Time!

1. False 2. True 3. True 4. True 5. False
6. True 7. False 8. False 9. True 10. False 11. True

Section 5.3
Are You Ready?
1. a. 3 terms b. -8 2. function, domain 3. -10

4. $10.2a^3$ 5. $17t^2u$ 6. $-3b^2 + b + 24$

Reading Time!
1. exponents, polynomial 2. polynomial in one variable 3. descending, leading, coefficient 4. one, polynomial in two variables, polynomial is three variables 5. ascending 6. monomial, binomial, trinomial 7. degree of a term, exponent, sum, 0, no defined degree 8. degree of the polynomial 9. linear, quadratic, cubic 10. evaluate, input, output 11. one, parabolas, smooth, continuous 12. like terms 13. combine like terms, same 14. add 15. subtract, change the signs 16. opposites, -1

Section 5.4
Are You Ready?
1. $90a$ 2. $10x - 15$ 3. $28y + 56$ 4. x^{14} 5. -30

6. $9x^2 + 3x + 2$

Reading Time!
1. D 2. I 3. C 4. F 5. H 6. J 7. A 8. E 9. K 10. B 11. G 12. M 13. L

Section 5.5
Are You Ready?
1. $2 \cdot 2 \cdot 3 \cdot 3 \cdot 3 = 2^2 \cdot 3^3$ 2. t^5 3. $-h + 9$ 4. 4

5. $6b^4 + 12b^2$ 6. $m^2 - 3m - 10$

Reading Time!
1. greatest, factor, GCF 2. coefficient, prime, variable, common, factors 3. 1 4. factoring out 5. prime, irreducible 6. opposite 7. factoring, grouping 8. Step 1: group, factor; Step 2: common, group; Step 3: common factor, regroup 9. multiplication, 1, 1 10. factor, completely 11. first 12. formula, variable, isolate, opposite

Section 5.6
Are You Ready?
1. 1 2. $x^2 + 2x - 48$ 3. $4a^2 + 36a + 81$ 4. 2 5. 2 and 5 6. -3 and 6

Reading Time!
1. perfect-square trinomials 2. $(A + B)^2$ 3. $(A - B)^2$ 4. coefficient of the squared variable 5. two binomials 6. two integers whose product is c and sum is b 7. same sign 8. different signs 9. trial-and-check method 10. Step 1, Step 5, Step 2, Step 4, Step 3 11. substitution 12. key number 13. Step 4, Step 2, Step 5, Step 3, Step 1

Section 5.7
Are You Ready?
1. $n^2 - 81$ 2. $64d^2$ 3. a. 27 b. 216 4. $64a^3$

5. $b^3 + 64$ 6. $8a^3 - 1$

Reading Time!
1. L 2. H 3. F 4. A 5. E 6. C 7. J 8. K 9. B 10. I

Section 5.8
Are You Ready?
1. a. 3 b. 6 2. Yes, both do. 3. $3a^3 - 27a^2 - 210a$

4. $8mn^2$

Reading Time!
1. product, polynomials 2. factor, factor, GCF, positive 3. terms, difference, sum, cubes, perfect-square, trial-and-check, grouping, four, grouping 4. factors, completely 5. check, multiplying

Section 5.9
Are You Ready?
1. 0 2. -4 3. 0 4. $(x - 3)(x + 2)$

5. $(3n + 2)(n - 1)$ 6. $(x + 11)(x - 11)(x - 1)$

Reading Time!
1. polynomial equation 2. $ax^2 + bx + c = 0$, a, b, c are real numbers, $a \ne 0$ 3. solution of a quadratic equation 4. Zero-Factor Property 5. Write the equation in standard form. 6. Factor the polynomial. 7. Use the zero-factor property to set each factor equal to zero. 8. Solve each resulting equation. 9. Check the results in the original equation. 10. extension of the zero-factor property 11. three solutions 12. should be discarded or rejected

Section 6.1
Are You Ready?
1. a. 0 b. undefined c. -1 2. $\frac{9}{7}$ 3. $4a^4(3a - 4)$

4. $(a + 10)(a - 10)$ 5. $(5n - 2)(n + 5)$

6. $(2x + 1)(x^2 - 7)$

Reading Time!
1. rational expression 2. rational function 2. evaluate an expression 3. $c(2) = \frac{1.50(2) + 6}{2} = 4.5$ 4. domain 5. reciprocal function 6. asymptote 7. vertical asymptote 8. horizontal asymptote 9. $\frac{a}{b} = \frac{c}{d}$ if and only if $ad = bc$; $\frac{a}{1} = a$ and $\frac{a}{a} = 1$; $\frac{ak}{bk} = \frac{a}{b} \cdot \frac{k}{k} = \frac{a}{b}$; $-\frac{a}{b} = \frac{-a}{b} = \frac{a}{-b}$ 10. first step to simplify rational expressions 11. simplify a rational expression 12. equivalent expressions 13. -1 14. opposites

Pre-Class Prep Answer Section

Section 6.2
Are You Ready?
1. $\frac{3}{56}$ 2. $\frac{3}{2}$ 3. $\frac{9}{5}$ 4. $x^3(1-x)$ 5. $\frac{x+5}{x-5}$ 6. $\frac{x^2+5x+25}{2}$
Reading Time!
1. J 2. F 3. K 4. L 5. H 6. C 7. I 8. A

Section 6.3
Are You Ready?
1. $\frac{8}{11}$ 2. $\frac{7}{15}$ 3. x^2+3x+4 4. 1 5. $\frac{2}{a-2}$

6. $5(x+1)(x-1)$
Reading Time!
1. True 2. False 3. True 4. False 5. True 6. True
7. False 8. False 9. True 10. True

Section 6.4
Are You Ready?
1. division 2. $\frac{7}{28}$ 3. $(a-2)(a+1)$ 4. $12a^2-20$
Reading Time!
1. complex, complex fraction 2. Division; *Step 1*:
numerator, single, single; *Step 2*: division, reciprocal;

Step 3: simplify 3. $\frac{3a}{b}\cdot\frac{b^2}{6ac}$ 4. LCD; *Step 1*: LCD,

all; *Step 2*: 1; *Step 3*: no; *Step 4*: simplify 5. x

Section 6.5
Are You Ready?
1. a. $\frac{2}{5}$ b. $2a^5$ 2. $8x^5y^5-20x^4y^7+2xy^2$ 3. 36

4. $2x^2$
Reading Time!
1. divide, simplifying fractions, quotient rule

2. positive 3. monomial, term 4. $\frac{A}{D}+\frac{B}{D}$ 5. common

6. long division 7. divisor, quotient, remainder,
dividend 8. long-division, divide, subtract
9. *Step 1*: descending; *Step 2*: less, divisor, quotient,
remainder 10. missing, placeholder, 0

Section 6.6
Are You Ready?
1. $3x+4$ 2. $2x+5+\frac{1}{x-2}$ 3. 26 4. $27x^2$
Reading Time!
1. E 2. I 3. B 4. G 5. D 6. J 7. A 8. F

Section 6.7
Are You Ready?
1. $30x$ 2. $8x$ 3. $-6, 9$ 4. 4
Reading Time!
1. rational equation 2. solve a rational equation
3. Step 1, Step 6, Step 3, Step 2, Step 4, Step 5
4. extraneous solution 5. fraction-clearing method
6. circle the variable you are solving for in the formula
7. multiply both sides of the formula by the LCD

Section 6.8
Are You Ready?
1. $t=\frac{d}{r}$ 2. $\frac{x}{5}$ 3. $20x$ 4. 18 5. $4\frac{4}{9}$ 6. 25, 4
Reading Time!
1. D 2. C 3. F 4. A 5. G 6. B 7. E 8. H

Section 6.9
Are You Ready?
1. a. ad b. bc 2. 27 3. $-2, 9$ 4. 5 5. 1,620 6. $90°$
Reading Time!
1. ratio 2. the three ways to write a ratio 3. rate
4. unit cost 5. proportion 6. extremes 7. means
8. The Fundamental Property of Proportions 9. cross
products 10. solving the proportion 11. similar
12. proportion 13. $y=kx$ for some nonzero constant k
14. constant of variation or constant of proportionality

15. Step 3, Step 1, Step 4, Step 2 16. $y=\frac{k}{x}$ for some

nonzero constant k 17. joint variation 18. $y=kxz$ for
some nonzero constant k 19. combined variation

Section 7.1
Are You Ready?
1. a. 225 b. 64 2. a. $\frac{49}{25}$ b. 0.04 3. a. -216 b. 81

4. a. a^8 b. x^9 5. $x^2+16x+64$ 6. 5
Reading Time!

1. square root 2. reverse 3. $\sqrt{}$, principal

4. $\sqrt{a}$, $-\sqrt{a}$, principal 5. radicand, radical 6. radical
expression 7. real numbers, imaginary numbers
8. rational, irrational, real number 9. $|x|$, principal,
absolute value 10. square root, radical
11. nonnegative 12. pendulum 13. cube root

14. reverse 15. $\sqrt[3]{a}$, index, radicand, radical
16. perfect cube, x 17. cube root, radical

18. $\sqrt[n]{a}$, index 19. odd, even 20. x, $|x|$

21. a. x b. 0 c. x, not a real number

Section 7.2
Are You Ready?

1. a. 8 b. –4 2. a. 3 b. $\frac{1}{2}$ 3. a. $|x+5|$ b. $3a^2$

4. 216 5. $\frac{1}{49}$ 6. x^9

Reading Time!

1. definition of $x^{1/n}$ 2. not a real number

3. a. x b. 0 c. $x < 0$ 4. definition of $x^{m/n}$

5. radical form 6. $(5xyz)^{1/2}$ 7. definition of $x^{-m/n}$

8. $x^{-m/n}$ 9. a^4b^3 10. absolute value symbols
11. Step 2 12. Step 3 13. Step 1

Section 7.3
Are You Ready?

1. a. 4 b. = 2. a. –2 b. 2 3. $27a^4b^8$ 4. $7x, 9x$

5. $14m^2 - m$ 6. $\frac{5}{6}$

Reading Time!

1. True 2. False 3. False 4. True 5. True 6. True
7. True 8. False 9. True 10. False 11. True 12. True

Section 7.4
Are You Ready?

1. $25a^{12}$ 2. $54t^6 + 18t^5$ 3. $2x^2 + x - 3$ 4. $1 - 2\sqrt{14}$

5. $x^2 - 100$ 6. $\frac{18}{27a}$

Reading Time!

1. D 2. H 3. F 4. J 5. I 6. M 7. A 8. C 9. L 10. E
11. B 12. K

Section 7.5
Are You Ready?

1. a. $x - 1$ b. $x^3 + 7$ 2. $18x + 45$ 3. $x^2 - 8x + 16$

4. –3, 9 5. $x^3 + 6x^2 + 12x + 8$ 6. $4 - 4\sqrt{x} + x$

Reading Time!

1. radical equation 2. equal, equal, $x^n = y^n$

3. nonnegative, $\left(\sqrt{a}\right)^2$, a 4. solutions, check

5. original, extraneous 6. solving, isolate, both, same,
index, radical, solve, check 7. real number, a 8. two,
left, right 9. radical, power rule 10. variable, isolate

Section 7.6
Are You Ready?

1. 180° 2. $a\sqrt{3}$ 3. $\frac{25\sqrt{3}}{3}$ 4. 13 5. 3.46 6. $\frac{14\sqrt{3}}{3}$

Reading Time!

1. K 2. J 3. F 4. E 5. G 6. B 7. H 8. A 9. I 10. C

Section 7.7
Are You Ready?

1. No real number squared is equal to –16.

2. $6x^2 - 2x$ 3. $14n^2 + 19n - 3$ 4. $\sqrt{63}$

5. $\frac{7\left(\sqrt{x} - 4\right)}{x - 4}$ 6. 21 R3

Reading Time!

1. complex number system 2. $i = \sqrt{-1}$ 3. –1 4. $3i$
5. $i\sqrt{b}$ 6. complex number 7. imaginary numbers
8. real part 9. imaginary part 10. adding and
subtracting complex numbers 11. addition of complex
numbers 12. subtraction of complex numbers 13.
multiplying complex numbers 14. $-10 + 0i$ 15.
complex conjugates 16. division of complex numbers

17. i^R

Section 8.1
Are You Ready?

1. a. $2\sqrt{7}$ b. $6i$ 2. $\frac{\sqrt{30}}{6}$ 3. $\frac{9}{2}$ 4. $\frac{49}{4}$ 5. 169

6. $(x - 4)^2$

Reading Time!

1. square, $x^2 = c$, $x = -\sqrt{c}$ 2. double-sign 3. real
numbers 4. square root property, binomial
5. completing, square 6. square, one-half,

$x^2 + bx + \left(\frac{1}{2}b\right)^2$ 7. *Step 1*: coefficient, dividing; *Step*

2: variable, constants; *Step 3*: complete, one-half,
squaring, adding; *Step 4*: perfect-square trinomial;
Step 5: square root property; *Step 6*: check, original
8. leading coefficient 9. solutions, complex numbers

Section 8.2
Are You Ready?

1. 11 2. $3\sqrt{5}$ 3. 3; 2, –1, 7 4. $\frac{3}{4}$, –2 5. not a real
number 6. 3.87
Reading Time!
1. F 2. H 3. B 4. E 5. G 6. I 7. C 8. A

Section 8.3
Are You Ready?

1. 84 2. a. 2 b. $\frac{2}{3}$ 3. a. x b. a^2 4. $\pm 2i$

5. a. 9 b. $-\frac{1}{8}$ 6. 5

Pre-Class Prep Answer Section

Reading Time!

1. discriminant 2. positive 3. 0 4. negative 5. a perfect square 6. positive and not a perfect square
7. square root property 8. completing the square
9. factoring and zero-factor property 10. quadratic formula 11. quadratic in form 12. $\sqrt{x}$ 13. shared-work problem

Section 8.4
Are You Ready?

1. 2. -2

3. $x^2 + 8x + 16 = (x+4)^2$ 4. $x^2 + x + \frac{1}{4} = \left(x + \frac{1}{2}\right)^2$

5. A repeated solution of -3 6. A repeated solution of -2 7. -5 8. A vertical line

Reading Time!

1. I 2. M 3. F 4. A 5. N 6. O 7. G 8. L 9. C
10. M 11. B 12. E 13. J 14. H 15. D

Section 8.5
Are You Ready?

1. yes 2. $-5, 10$
3.

4.

5. $-3, 3$
6.

Reading Time!

1. quadratic, $ax^2 + bx + c \le 0$ 2. *Step 1*: standard form, related; *Step 2*: critical numbers, number line;
Step 3: interval, test value, includes, true;
Step 4: endpoints, included 3. interval testing
4. *Step 1*: standard form, related; *Step 2*: denominator, zero; *Step 3*: critical numbers, number line; *Step 4*: interval, test value, includes, true; *Step 5*: endpoints, included, exclude 5. *Step 1*: boundary line, allows, solid, not allowed, dashed; *Step 2*: test point, origin, original, satisfied, contains, not satisfied, other
6. nonlinear

Section 9.1
Are You Ready?

1. $9x^2 + 2x - 8$ 2. $72x^3 - 45x^2 - 32x + 20$
3. $2x^3 - x^2$ 4. $x + 2$

Reading Time!

1. algebra of functions 2. $f(x) + g(x)$
3. $f(x) - g(x)$ 4. $f(x) \cdot g(x)$ 5. $\frac{f(x)}{g(x)}$
6. compositions of functions 7. $f(g(x))$
8. composite function 9. functions f and g such that $h(x) = (f \circ g)(x)$ 10. $(C \circ F)(t)$

Section 9.2
Are You Ready?

1. D: $\{-2, 3, 5, 9\}$; R: $\{-3, 1, 8, 10\}$ 2. Yes 3. function
4. $16x + 6$

Reading Time!

1. inverse 2. one-to-one 3. horizontal, one-to-one, once 4. one-to-one, inverse, f^{-1}, (y, x) 5. *Step 1*: notation, y; *Step 2*: interchange; *Step 3*: solve; *Step 4*: substitute 6. inverse, x, $\left(f^{-1} \circ f\right)(x)$ 7. inverses, mirror images, $y = x$

Section 9.3
Are You Ready?

1. a. 3^{12} b. 3^{32} 2. a. 8 b. 1 c. $\frac{1}{4}$ 3. a. $\frac{1}{9}$ b. 1 c. 27
4. line, parabola

Reading Time!

1. D 2. H 3. C 4. J 5. F 6. I 7. K 8. G 9. E
10. L 11. M 12. A 13. N 14. B

Section 9.4
Are You Ready?

1. 0, 1, 2 2. a. 125 b. 3 3. a. 1 b. $\frac{1}{2}$ 4. a. 1,000
b. $\frac{1}{100}$

Reading Time!

1. True 2. True 3. True 4. False 5. False 6. True
7. True 8. False 9. True 10. True 11. True 12. False
13. False 14. True 15. True

Section 9.5
Are You Ready?

1. 2.25 2. 1 3. 2.7 4. 82.05 5. $3^6 = x$ 6. $(8, 3)$

Reading Time!
1. natural, domain, range   2. horizontal, x-axis, $(0, 1)$,
$(1, e)$ 3. translated   4. $A = Pe^{rt}$, annual growth rate,
decrease   5. semiannually, quarterly, continuously
6. natural, $\ln x$ 7. exponent, $\ln x$, $e^{\ln x}$ 8. power
9. logarithmic, exponential   10. $f(x) = \ln x$,
$y = \ln x$ 11. $(0, \infty), (-\infty, \infty)$ 12. double
13. continuously, $\frac{\ln 2}{r}$

Section 9.6
Are You Ready?

1. 3   2. 3   3. –9   4. 2   5. $x^{\frac{1}{2}}$, $\sqrt{x-2}$ 6. 1.6094
Reading Time!
1. $\log_b 1 = 0$ 2. $\log_b b = 1$ 3. $\log_b b^x = x$
4. $b^{\log_b x} = x$ $(x > 0)$ 5. $\log_b M + \log_b N$
6. $\log_b M - \log_b N$ 7. $p \log_b M$ 8. expand a
logarithmic expression   9. condense logarithmic
expressions   10. change-of-base formula   11. $\frac{\log_a x}{\log_a b}$
12. hydrogen ion   13. pH scale   14. $-\log\left[H^+\right]$

Section 9.7
Are You Ready?
1. a. 4 b. –3   2. 1.1309   3. $x \log_3 8$ 4. 1   5. a. $\log_2 5x$
b. $\ln \frac{10}{2t+1}$ 6. undefined
Reading Time!
1. D   2. H   3. C   4. J   5. F   6. A   7. K   8. B   9. G
10. E   11. M   12. L   13. I

Section 10.1
Are You Ready?
1. $x^2 - 16x + 64$ 2. $x^2 - 6x + 9 = (x-3)^2$
3. $(y+5)^2$ 4. $-3(y^2 + 4y)$

Reading Time!
1. intersection, circles, parabolas, ellipses, hyperbolas
2. focus, paraboloid   3. points, center, radius
4. standard, (h, k), $(x-h)^2 + (y-k)^2 = r^2$ 5. center,
radius   6. $x^2 + y^2 + Dx + Ey + F = 0$ 7. distance,
$d = \sqrt{(x_2 - x_1)^2 + (y_2 - y_1)^2}$ 8. equidistant, focus,
directrix
9. $y = a(x-h)^2 + k$, (h, k), $x = h, a > 0, a < 0$
10. $y = ax^2 + bx + c$, $x = ay^2 + by + c$
11. $x = a(y-k)^2 + h$

Section 10.2
Are You Ready?
1. ± 9 2. $y + 4$ 3. $4x^2 + y^2$ 4. $2\sqrt{2}$
Reading Time!
1. D   2. H   3. C   4. F   5. B   6. G   7. E   8. I   9. A

Section 10.3
Are You Ready?
1. ± 4 2. $x + 2$ 3. $\frac{2}{3}$ 4. –10
Reading Time!
1. hyperbola   2. foci   3. center   4. vertices
5. central rectangle   6. Asymptotes   7. $\frac{x^2}{a^2} - \frac{y^2}{b^2} = 1$
8. $\frac{y^2}{a^2} - \frac{x^2}{b^2} = 1$ 9. $\frac{(x-h)^2}{a^2} - \frac{(y-k)^2}{b^2} = 1$
10. $\frac{(y-k)^2}{a^2} - \frac{(x-h)^2}{b^2} = 1$ 11. hyperbola
12. discovery of the atomic structure of matter
13. $4x^2 - y^2 = 16$ 14. hyperbola   15. circle
16. ellipse   17. parabola

Section 10.4
Are You Ready?
1.   2. $-1, \frac{5}{9}$ 3. $(4, 12)$

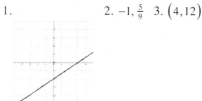

4. a. $\pm \frac{\sqrt{5}}{3}$ b. $\pm 3\sqrt{2}$

Reading Time!
1. True 2. False 3. False 4. True 5. False 6. True

Section 11.1
Are You Ready?
1. 1, 4, 6, 4, 1   2. $a^2 + 2ab + b^2$ 3. 120   4. 30
Reading Time!
1. expand, sum   2. Pascal's, sum, above   3. $n!$,
product, $n(n-1)(n-2)(n-3) \cdot \cdots \cdot 3 \cdot 2 \cdot 1$ 4. 1

5. $n!$ 6. $(a+b)^n$ 7. binomial, sum, decrease,

increase   8. $\dfrac{n!}{r!(n-r)!} a^{n-r} b^r$

Section 11.2
Are You Ready?
1. 21   2. a. 1 b. −1   3. 8   4. 52
Reading Time!
1. F   2. C   3. H   4. J   5. A   6. I   7. D   8. M   9. B
10. K   11. E   12. L   13. G

Section 11.3
Are You Ready?
1. 48   2. a. 625 b. $\frac{1}{27}$ 3. $\frac{1}{4}$ 4. ±12
Reading Time!
1. geometric series   2. $a_n = a_1 r^{n-1}$ 3. first term

4. $r = \dfrac{a_{n+1}}{a_n}$ 5. geometric means   6. $S_n = \dfrac{a_1 - a_1 r^n}{1-r}$,

where $r \neq 1$ 7. geometric series   8. infinite geometric

series   9. partial sum   10. $S = \dfrac{a_1}{1-r}$ 11. use a

geometric sequence   12. find the sum of the terms of
the infinite geometric sequence

Appendix B: Analyzing Your Test Results

Once your graded test is returned to you, use these guidelines to help you learn from this experience.

- ☐ Refer to the following list, and classify each error that you made on the test. Then compile the results to determine what type of errors you make most often.
  - o Sign error
  - o Arithmetic error
  - o Misunderstanding (or lack of understanding) of vocabulary
  - o Misuse of an algebraic property or rule
  - o Didn't follow directions
  - o Copying error
  - o Graphing error
  - o Ran out of time

- ☐ If you need further help in identifying errors in your solutions, review your test with a classmate, a tutor, or your instructor.

- ☐ Always double-check your test score to make sure there were no errors in the grading.

- ☐ Take clear notes on the solutions when reviewing the test in class.

- ☐ Keep your graded tests in this binder so that you can use them to prepare for the final exam.

- ☐ Check your syllabus or ask the instructor if there will be a make-up exam or any extra credit offered that might help to improve your grade.

- ☐ If you did well, pat yourself on the back and celebrate!